Notes on Meteorology

Notes on Meteorology

Kemp & Young

STANFORD MARITIME LONDON

Stanford Maritime Limited
Member Company of the George Philip Group
12–14 Long Acre London WC2E 9LP

First Published 1961
Second edition 1966
Third edition 1971
Reprinted 1972, 1974, 1977, 1980, 1984, 1987

Printed and bound in Great Britain
at The Bath Press, Avon

ISBN 0 540 07334 2

PREFACE TO FIRST EDITION

Weather conditions are ever changing and, in order to make the best use of favourable conditions or to counter unfavourable ones successfully, the seafarer should have some theoretical as well as practical knowledge of meteorology.

Many of the theories of weather phenomena are revised from time to time and, so as to keep the text concise, we have described the more common of these theories as simply as possible.

A large number of diagrams have been included to supplement the text and to enable the student to make his own forecast of the likely weather in the various situations.

Although written primarily to cover the needs of those studying for their examinations, all interested in meteorology, particularly yachtsmen, will find the book of great value.

Our thanks are due to Mr. P.N. Colepeper, F.R.Met.S., Master Mariner and First Class Air Navigator, for his painstaking reading of the text and for his helpful suggestions.

We are indebted to the Controller of Her Majesty's Stationery Office for permission to reproduce certain drawings of instruments from the Admiralty Manual of Seamanship and the map of the weather forecast areas supplied by the Meteorological Office.

KENLEY, SURREY
May 1961

J.F. KEMP
PETER YOUNG

PREFACE TO REVISED EDITION

The changeover to metrication with SI units has necessitated several alterations in the numerical quantities in this book but, wherever appropriate, both SI and Imperial values are given. It should be noted, however, that the imperial units are not necessarily an exact conversion from SI units.

The opportunity has also been taken to revise the text and introduce notes on further topics such as facsimile plotting and weather routeing so that the book may continue to fulfil its original purpose of providing a basic text on meteorology for examination candidates, yachtsmen and all interested in the subject of weather.

Our thanks are due to those who have read and criticised the revised text and to Negretti and Zambra (Aviation) Limited, who have supplied illustrations.

January 1971

J.F. KEMP
PETER YOUNG

CONTENTS

I	INSTRUMENTS	3
II	THE ATMOSPHERE	15
III	CLOUD AND PRECIPITATION	23
IV	WIND	33
V	ISOBARIC SYSTEMS	53
VI	TROPICAL STORMS	71
VII	CURRENTS AND ICE	78
VIII	WEATHER CHARTS AND ROUTEING	82
	INDEX	87

I
Instruments

In order to get as complete a picture as possible of the weather, careful observations should be made of the individual phenomena which go to make up the weather. Many of these observations are made visually; for example, the form of clouds, and direction of the wind. Others must be made by instruments; for instance one cannot 'feel' the pressure or the relative humidity although one may hazard a guess at the air temperature.

Various instruments have been designed to observe the different phenomena. The principal ones measure pressure, temperature and wind velocity, whilst others have been designed to measure sunshine hours and rainfall.

There is no necessity for a ship to carry every instrument, as a barometer and a hygrometer together with the broadcast weather information will enable one to make an accurate forecast for the next few hours. The Meteorological Office recognises this and usually supplies a precision aneroid or a mercurial barometer, a barograph, a sea thermometer and a hygrometer to observing ships.

PRESSURE AND ITS MEASUREMENT

Pressure is force per unit area. The unit of pressure in the SI system is the BAR which is approximately equivalent to a load of 10 tonnes weight per square metre ($10tf/m^2$).

The average pressure at ground level is slightly in excess of 1 Bar and to avoid the necessity of having large numbers of figures after the decimal point in order to express pressure accurately, the bar is divided into 1000 parts, each of these parts is a MILLIBAR, and pressure is expressed in these units.

For many years pressure was also expressed in millibars under the Imperial system although it was also, and still may be, expressed as "inches of mercury". This is the length of a column of mercury which will balance a column of air. As the column of air exerts a pressure of approximately 14.7 lbs per square inch the height of the column of mercury (relative density 13.6) necessary to exert this pressure is about 30 ins.

ISOBAR — A line joining places having equal pressure. On weather charts these are plotted at 4 millibar intervals, so that the isobars shown are divisible by 4.

ISALLOBAR. — A line joining places having an equal change of pressure. A study of these can give an indication of the direction of movement of pressure systems.

PRESSURE GRADIENT. This is the difference in pressure in unit distance measured at right angles to the isobars.

There is a DIURNAL RANGE of barometric pressure which results in the barometric pressure being higher than normal at 1000 and 2200 local time and lower than normal at 0400 and 1600 local time. The semi-diurnal pressure wave is due to the atmospheric tides which are caused by the sun and moon. It is possible that there are other causes as this semi-diurnal change of pressure is still being investigated.

The diurnal range is most marked in the tropics where the barometer is frequently 1.5 millibars above or below normal at the times mentioned above. It is less noticeable in higher latitudes where frequent pressure changes occur due to the passage of depressions. However, when the pressure gradient is constant and small, this daily range may be seen clearly on a barograph trace although the variation is very much less than in low latitudes.

MERCURIAL BAROMETER

This is constructed by filling a tube, about 1 metre (36 ins.) long with mercury. The end of the tube is temporarily closed and is inverted and placed into a reservoir of mercury. When the closure is removed it will be seen that the level of mercury falls in the tube. The space above the mercury at the top of the tube is known as a Torricellian vacuum (after Torricelli). If an air bubble were to get into this space it would depress the mercury (as the vacuum would no longer be complete) and an incorrect reading would result. To prevent this, an air trap is incorporated in the tube. A further refinement in the KEW PATTERN MARINE BAROMETER is the capillary tube between the air trap and the marine tube.

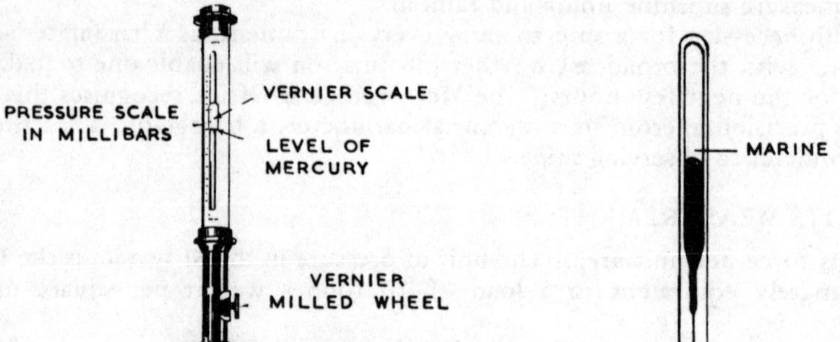

The mercurial barometer is liable to error on account of the following :—

1. CAPILLARITY. The surface tension of the mercury forms a MENISCUS and readings should always be taken at the top of this.
2. CAPACITY. The height of the barometer should be taken from the top of the mercury in the cistern to the top of the mercury in the marine tube. If the pressure increases, the level of the mercury in the cistern falls, so that the measurements cannot be taken from a fixed point. This error is compensated by adjusting the distance between the graduations. On the inch barometer the barometer inch will be seen to be 24/25ths of a linear inch. (In the FORTIN barometer used in laboratory work the level of mercury in the cistern is adjustable so that the readings can always be taken from the same level).
3. PUMPING. Due to the constant change in height above mean sea level of a barometer on a vessel in a seaway, there will tend to be a continual change of reading. This movement of the mercury will make it difficult to get an accurate reading. Gusty winds can also cause pumping. The effect of pumping is considerably reduced by fitting the capillary tube above the air trap. (If pumping is present try to get a mean of the highest and lowest readings).
4. HEIGHT. All readings should be corrected to sea level. Increase of height means a decrease of pressure by approximately 1 millibar for every 10m (30ft.). This correction can be made by tables or by the Gold Slide.
5. LATITUDE. Due to the earth being somewhat "flattened" at the poles, mercury weighs more at the poles than at the equator. For equal atmospheric pressures the barometer would appear to read less at the poles and more at the equator. Readings should all be corrected for a mean latitude of 45°. The correction can be made by tables or by the Gold Slide.
6. TEMPERATURE. The column of mercury will expand with an increase of temperature and contract with a decrease of temperature in exactly the same way as does a thermometer. All readings must be reduced to a standard temperature which is 285°K in the case of most millibar barometers and 28.6°F in the case of inch barometers. The correction can be made by tables or by the Gold Slide. N.B. The attached thermometer should always be read before the barometer as otherwise heat from the observer's body may give a false reading.
   The temperature at which a barometer reads correctly is known as the FIDUCIAL TEMPERATURE. In lat. 45° at sea level this is the same as the standard temperature, but at sea level in lat. 57° the fiducial temperature would be 291°K, and in lat. 21° at sea level it would be 273°K. Whereas at 20m (60ft.)above sea level in 45°it would be 297°K. (All the foregoing figures are to the nearest degree).
7. OBSERVATIONAL ERRORS.
   (a) the barometer should always be upright; it is the vertical height of the column of mercury that balances the column of air. If the barometer is not upright, too high a reading is obtained.
   (b) the back and front of the vernier must be on the same level as the observer's eye, otherwise the reading will be too high.

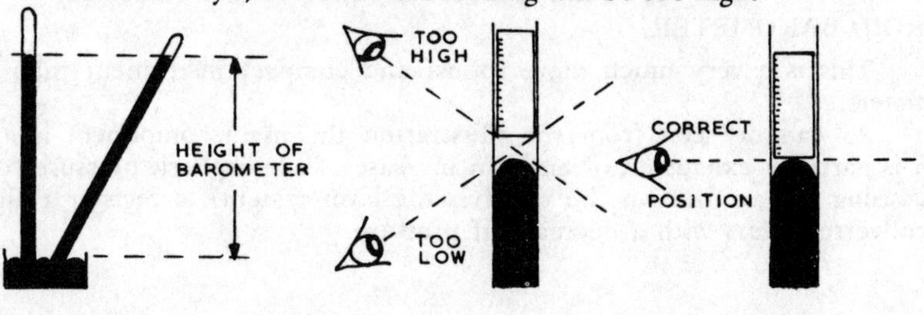

GOLD SLIDE

This takes its name from its inventor Lt. Col. Gold, and gives a rapid means of getting the latitude, height and temperature correction.

To use the slide set the height of the barometer above sea level against the latitude and read off the correction opposite the top of the mercury in the thermometer.

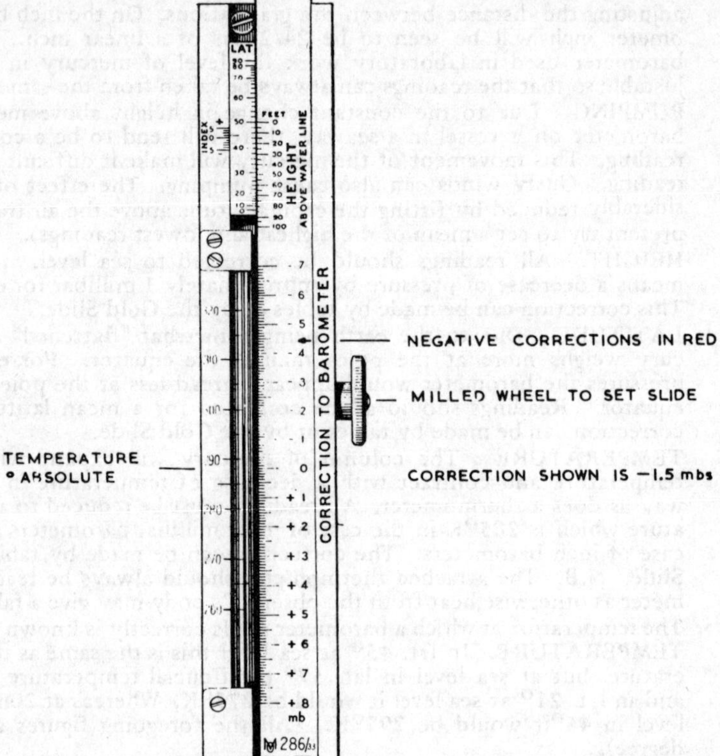

ANEROID BAROMETER.

This is a very much more robust and compact instrument than the mercurial barometer.

As can be seen from the illustration the main component is a vacuum box which is partially exhausted of air. An increase of atmospheric pressure compresses this box causing the pointer on the dial (via the lever system) to register a higher pressure. The converse occurs with a decrease of pressure.

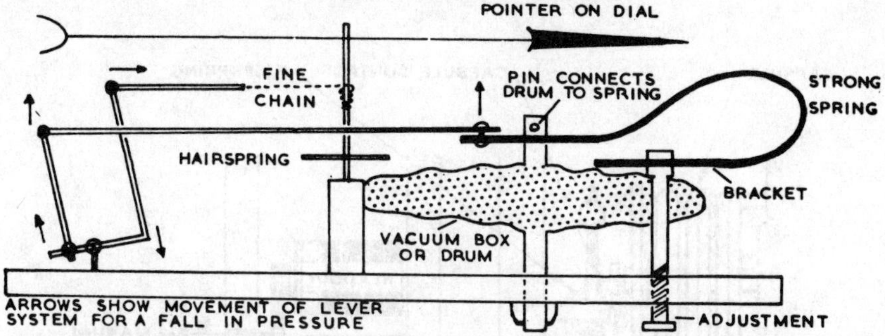

As there is no mercury in this barometer there are no corrections for latitude or temperature, but a height correction must be applied. There are no errors due to capillarity, capacity or pumping. There is an adjustment screw on the back of the instrument to take out any index error.

The greater the area of the vacuum box the greater the accuracy of the instrument. It is usual to give the barometer a light tap before reading; this helps to free the fine chain which may stick if pressure changes are only small.

PRECISION ANEROID BAROMETER

This instrument is now supplied, when available, to all observing ships in place of the Kew Pattern mercurial barometer. It is simpler to transport and to read whilst temperature correction is unnecessary. Height corrections can be 'built in' by resetting the datum on the instrument. A pressure choke can be attached if rapid height variations, leading to rapid pressure variations, are expected; this smoothes the variations to negligible amounts. The photograph shows the latest model (Mark 2) of this instrument.

The precision aneroid works on the same principle as that of any aneroid namely the movement of a "Pressure pile", here called a "capsule", caused by variations in air pressure. The difference between the precision aneroid and an ordinary aneroid is in the means of transmitting the change in air pressure to a reading. The diagram overleaf shows how this is done. This diagram and the photograph of the instrument have been kindly supplied by Negretti & Zambra (Aviation) Ltd., who manufacture the instrument.

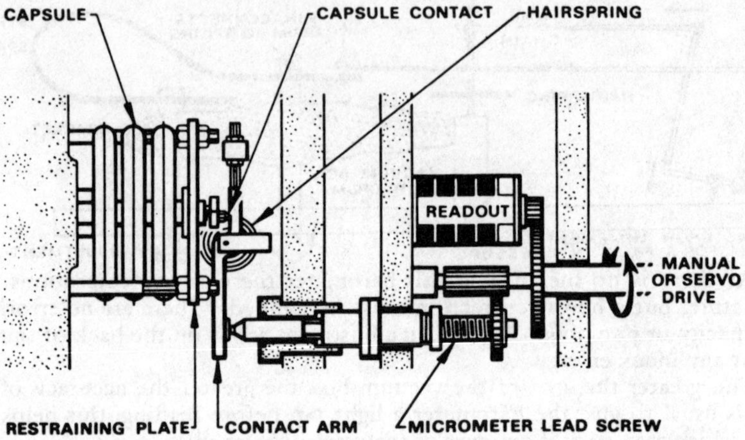

The movement of the' capsule' causes the capsule contact to move towards or away from the contact arm. If the contact is broken the 'magic eye' shows this and this is the instant at which to take the reading. The reading can range from 900mb to 1050mb and the accuracy to which it can be taken is 0.1mb. At the instant of taking the reading the capsule is only under atmospheric pressure as there is no load or pressure on it from the contact arm.

The only maintenance necessary is the infrequent renewal (about 3-6 months) of the small battery powering the electronic circuit (not illustrated here) which operates the minute cathode ray tube or "magic eye".

BAROGRAPH.

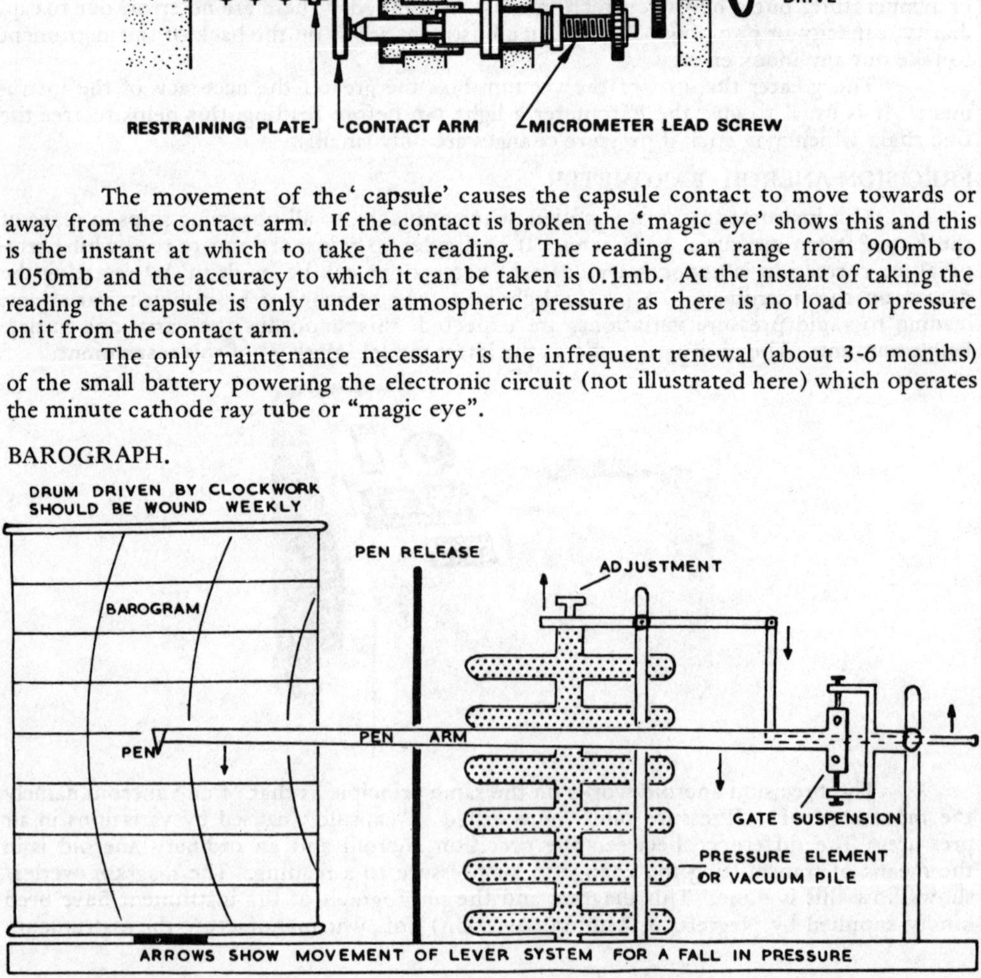

This is a recording aneroid barometer. As can be seen, the lever system connects the vacuum pile to a pen arm which makes a mark on the chart on the drum which is driven by clockwork. The drum is wound and the chart changed weekly.

The prime purpose of the barograph is to record the pressure tendency, which it would be impossible to observe with the mercurial or aneroid barometers unless an observer was detailed to record the pressure every five minutes or so.

TEMPERATURE AND ITS MEASUREMENT.

The forecasting of weather depends as much on a knowledge of temperature as of pressure.

The instruments for measuring temperature all depend on the expansion and contraction of liquids or metals when heated and cooled.

The thermometer, in its simplest form, consists of a capillary tube on the end of which is a bulb filled with mercury. This thermometer is graduated (as are all others) by placing it in pure melting ice and marking the position of the mercury, then placing it in boiling distilled water and again marking the position of the mercury. The barometric pressure in each case should be 760mm (30 ins.) of mercury. These two points are known as the fixed points of the thermometer. The part of the tube between these points is then divided into a number of equal divisions.

The number of divisions depends on the scale to be used. The various scales are shown below.

	Freezing Point Temperature	Boiling Point Temperature	Scale Divisions
Celsius (C)	0°	100°	100
Absolute or Kelvin (K)	273°	373°	100
Fahrenheit (F)	32°	212°	180
Réamur (R)	0°	80°	80

Conversions from one scale to another can be made quickly as :—

Celsius	=	(Fahrenheit^o - 32°) x 5/9
Fahrenheit	=	(Celsius^o x 9/5) + 32°
Absolute	=	Celsius^o + 273°
Réamur	=	Celsius^o x 4/5

It may be noted that at -40° the Celsius and fahrenheit scale readings are the same.

SPECIFIC HEAT of a substance is the number of Joules required to raise 1kg of the substance 1^oC. e.g. water is 4.182 whereas sand is 0.84, which means that a given quantity of sand will heat 5 times as much as the same quantity of water provided the same amount of heat is applied. It will also cool 5 times as quickly under similar conditions.

ISOTHERM. A line joining places having equal temperature.

AIR TEMPERATURE. The thermometer should be placed out of the direct rays of the sun and away from local draughts or warm air currents.

GROUND TEMPERATURE. This is taken at shore stations only, the thermometer being placed horizontally 50mm (2 inches) above the ground.

SEA TEMPERATURE. This may be taken with a thermometer, specially guarded against breakage, set inside a canvas bucket and trailed in the sea. However the more common method is to get a canvas bucket full of the surface water and push the thermometer into the bucket after it has been brought up on deck. In both cases the water must be from the surface and preferably well forward so as to be clear of all discharges.

DEW POINT TEMPERATURE is the temperature to which the air has to be cooled for the water vapour to condense out into water droplets. It is also known as the saturation temperature, and is dependent on the absolute humidity.

ABSOLUTE HUMIDITY is the actual weight of water vapour in a parcel of air and is expressed in grams per cubic metre. The greater the air temperature the more water vapour that it can absorb before becoming saturated.

RELATIVE HUMIDITY is the ratio between the amount of water vapour in the air and the amount that it can contain at that temperature. It is usually expressed as a percentage.

MAXIMUM THERMOMETER.

Usually used at shore stations in order to record the maximum daily temperature. As shown, they may be of two types, one having a constriction in the tube (as with a clinical thermometer) and the other with an index in the bore. With this latter type the mercury pushes the index up the tube when the temperature rises and when it falls the index is left in position. The maximum temperature is read at the end of the index nearest to the mercury. Mercury is used as it has a high boiling point.

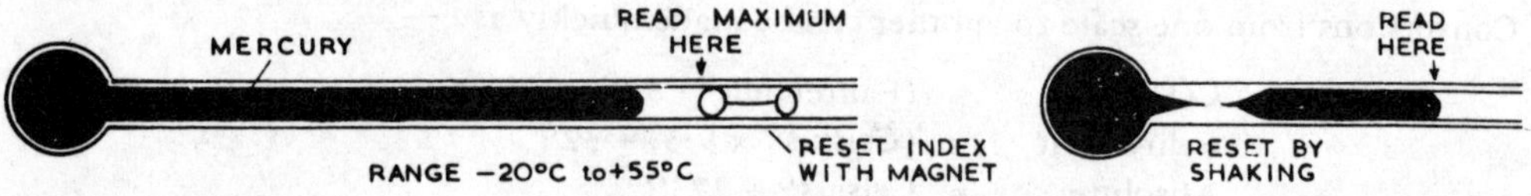

MINIMUM THERMOMETER.

As with the maximum thermometer, this is generally used at shore stations to record the minimum daily temperature. The liquid is usually alcohol as this has a low freezing temperature. The index is immersed in the alcohol and, as the temperature falls and the alcohol contracts, the surface tension of the alcohol draws the index down the tube. As the temperature increases the alcohol is free to flow past the index. The minimum temperature is read at the end of the index nearest the open end of the alcohol.

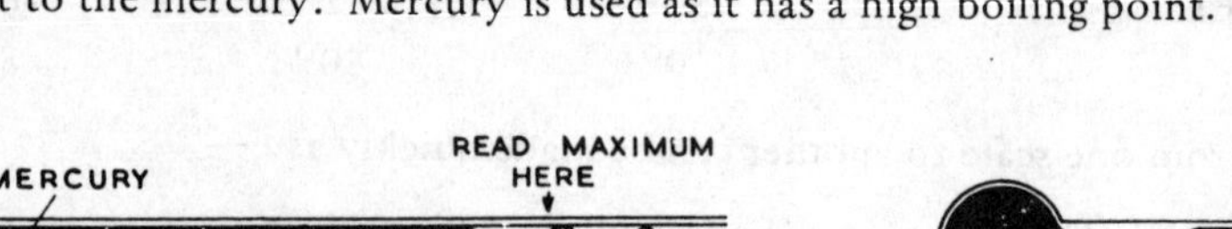

SIXES THERMOMETER.

This is a useful little instrument which incorporates a maximum and minimum thermometer and is much used by gardeners. The expansion of the alcohol in the round bulb as the temperature rises, forces the mercury round towards the pear shaped bulb, and in turn, forces the index up the tube. The converse occurs when the temperature falls. The maximum and minimum temperatures are read at the ends of the indices nearest the mercury.

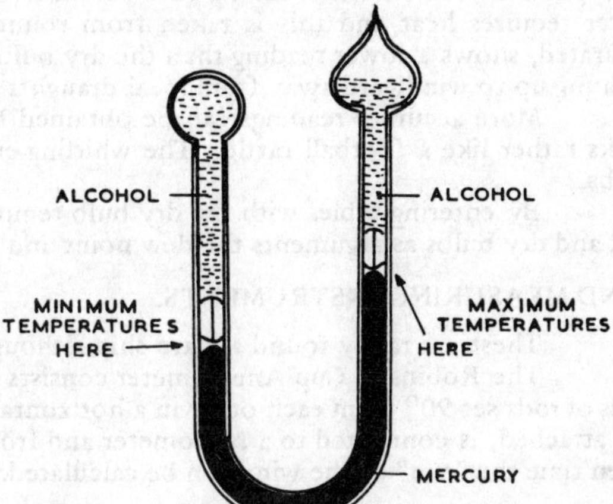

THERMOGRAPH.

This recording thermometer is not often seen aboard ship. A pen, attached to a metallic coil which expands and contracts, records the temperatures on a drum moved by clockwork.

PSYCHROMETER OR HYGROMETER.

Mason's Hygrometer consists of two thermometers mounted side by side in a Stevenson's screen. One is a dry bulb thermometer, the other a wet bulb thermometer.

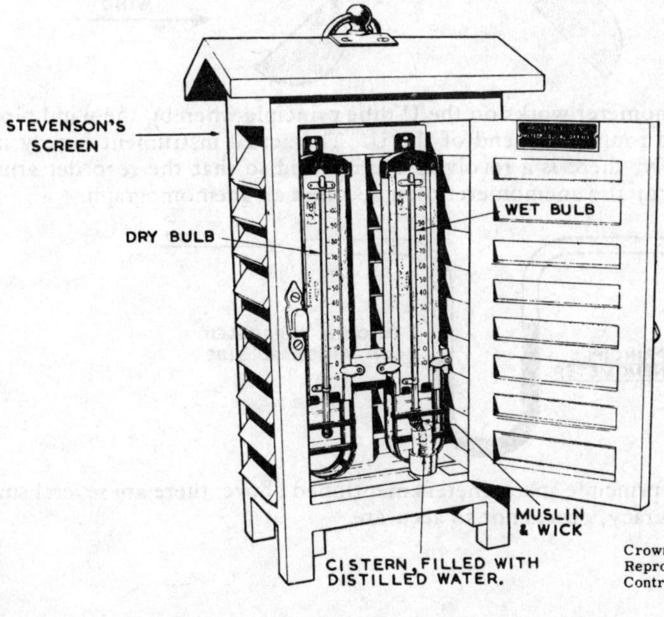

Cambric is wrapped round the bulb of the latter and it is kept moist by means of a piece of cotton wick leading to a container of distilled water. The evaporation of water requires heat and this is taken from round the wet bulb which, unless the air is saturated, shows a lower reading than the dry bulb. The screen and thermometers should be hung up to windward away from local draughts or warm air currents.

More accurate readings can be obtained by using a whirling psychrometer. This looks rather like a football rattle. The whirling ensures a steady flow of air over the two bulbs.

By entering tables with the dry bulb temperature and the difference between the wet and dry bulbs as arguments the dew point and relative humidity can be found.

WIND MEASURING INSTRUMENTS.

These are rarely found aboard ship although every shore station has one.

The Robinson Cup Anemometer consists of four hemispherical cups fixed to the ends of rods set 90° from each other in a horizontal plane. The spindle, to which the rods are attached, is connected to a tachometer and from the number of revolutions made in a given time the "run" of the wind can be calculated.

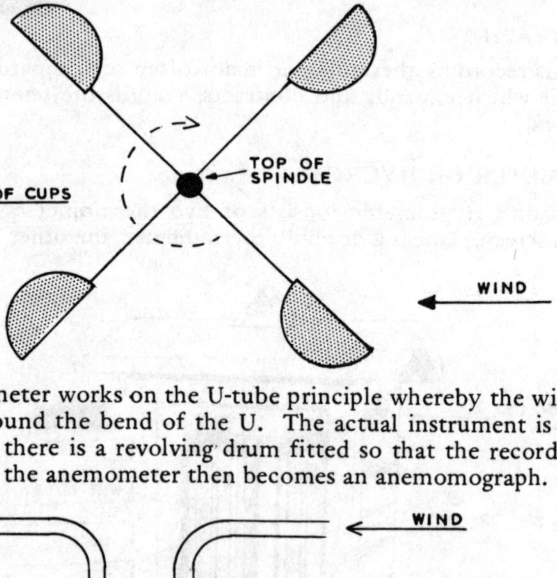

PLAN VIEW OF CUPS

The Dines anemometer works on the U-tube principle whereby the wind blowing in one side forces liquid round the bend of the U. The actual instrument is very much more refined, and usually, there is a revolving drum fitted so that the recorder arm can make a continuous record; the anemometer then becomes an anemomograph.

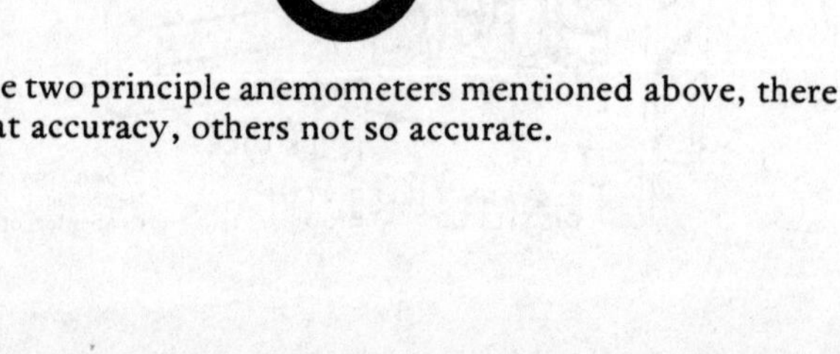

SHOWING PRINCIPLE OF DINES ANEMOMETER

Besides the two principle anemometers mentioned above, there are several smaller types, some of great accuracy, others not so accurate.

RAIN GAUGE.

This is infrequently found aboard ship but will be found at all shore stations.

It consists of a cylinder 130mm (5 ins.) in diameter having inside a small jug. The rain is collected in this jug by means of a funnel, which also prevents the rainfall evaporating.

To measure the rainfall the water is poured from the jug into a measuring glass which is graduated in millimetres or hundredths of an inch.

Snowfall is measured with a ruler and it is reckoned that 300mm (12 ins.) of snow is the equivalent of 25mm (1 in.) of rain.

ISOHYET. A line joining places having equal rainfall.

SUNSHINE RECORDER.

This is found at all shore stations and at most coastal resorts but not aboard ship.

In the Campbell-Stokes sunshine recorder, the sun's rays are focused by means of a glass sphere onto a piece of sensitized paper which discolours when the sun shines. The hours of sunshine can then be counted up.

ISOHEL. A line joining places having equal sunshine.

HYDROMETER.

Although hardly an instrument for foretelling weather in the usual sense, the hydrometer can none the less be valuable in helping to decide which currents may be influencing the vessel.

The hydrometer works on Archimedes' principle that a floating body displaces its own weight of the liquid in which it floats.

It consists of a float chamber through which passes a stem, the lower end of which is weighted so that it floats upright. The upper end is graduated to read the density of water. The instrument is frequently made of glass although polished steel hydrometers are more robust. Whichever type is used the surface should always be kept clean.

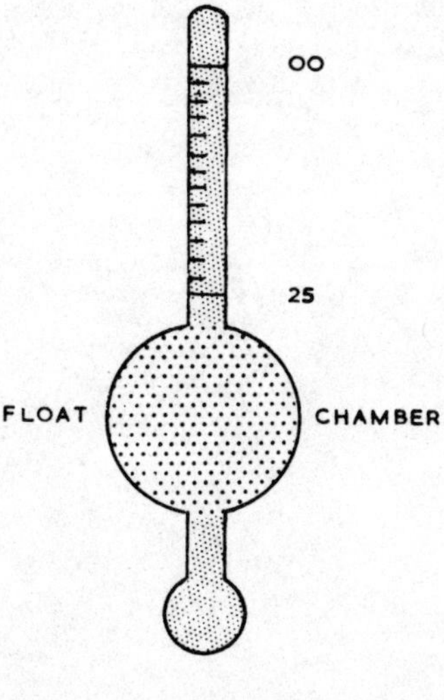

SALINOMETER.

This is a similar instrument to the hydrometer differing only in the graduations which are parts of salt per thousand (^{o}/oo).

ISOHALINE. A line joining places having equal salinity.

THE RADIO-SONDE.

This is a miniature radio station which is carried into the air by a large hydrogen filled baloon. Attached to the transmitter is an aneroid barometer, a thermometer and a hygrometer, and readings of these are radioed back to the point of release. This is frequently an ocean weather ship. The upper wind velocity and direction may also be found when the radio-sonde is tracked by radar.

NEPHOSCOPE.

This instrument is used to find the velocity/height ratio of cloud. If the height is known, the velocity is known, and vice-versa.

There are two main types; one, the BESSON'S comb nephoscope, looks rather like a garden rake and here the time of passage of a cloud between the spikes is noted.

In the FINEMAN reflecting nephoscope the time of passage of the cloud's reflection between rings on a horizontal mirror is noted.

ISONEPH. A line joining places having equal cloud amounts.

II
The Atmosphere

Most of the weather changes take place in the lower layer of the atmosphere which is known as the TROPOSPHERE. This layer extends about 11 miles high over the equatorial regions and about 5 miles high over the polar regions. The layer above the troposphere is known as the STRATOSPHERE where there is little water vapour and the lower part appears to be isothermal. There is a rise in temperature towards its upper limit. The boundary between the troposphere and the stratosphere is known as the TROPOPAUSE. About 20 miles above the earth the stratosphere gives way to the OZONOSPHERE, where there is a high concentration of ozone which absorbs ultra violet radiation. The IONOSPHERE starts about 50 miles above the earth and this is the layer that contains the various radio wave reflecting areas known as the KENNELLY-HEAVISIDE (55 miles) and APPLETON (150 miles) layers.

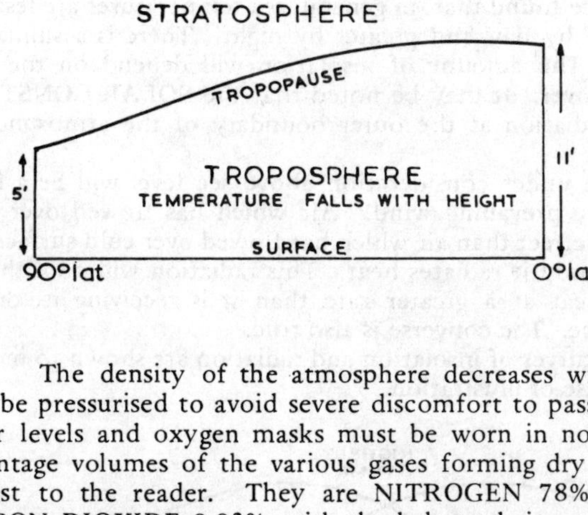

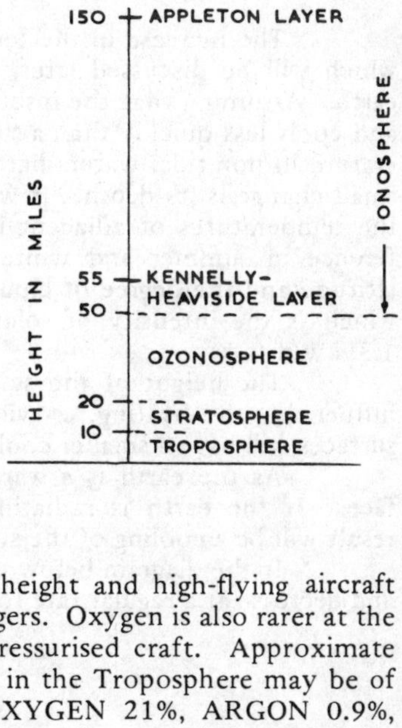

The density of the atmosphere decreases with height and high-flying aircraft must be pressurised to avoid severe discomfort to passengers. Oxygen is also rarer at the higher levels and oxygen masks must be worn in non-pressurised craft. Approximate percentage volumes of the various gases forming dry air in the Troposphere may be of interest to the reader. They are NITROGEN 78%, OXYGEN 21%, ARGON 0.9%, CARBON DIOXIDE 0.03%, with the balance being made up of traces of HYDROGEN, HELIUM, NEON, KRYPTON, RADON, XENON and OZONE.

As most of the weather changes take place in the troposphere, it is in this region that the changes in temperature are most important.

The sun, at a distance of about 93,000,000 miles and at a temperature of about 6,000°C, is the principal source of light and heat for the earth. The heat from the sun travels to the earth in the form of short wave radiation, which passes through the atmosphere without appreciably warming it. On striking the earth some of the heat will be absorbed to warm the earth. The heat received at the earth from the sun is known as INSOLATION.

The amount of insolation per unit area varies with latitude. It can be seen in the sketch below that a band of rays has to heat a very much larger area in a high latitude than it does in a low latitude.

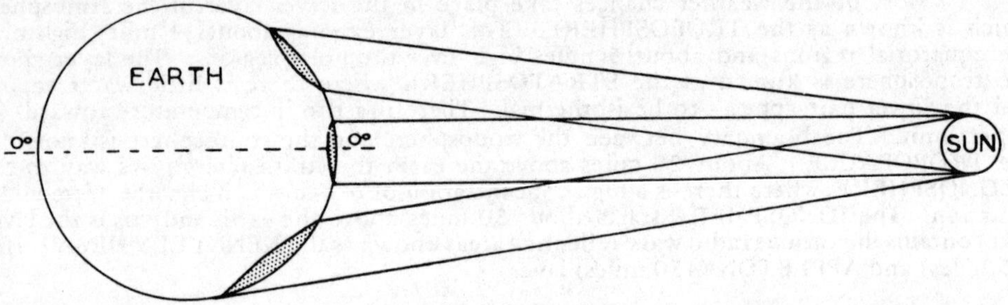

The increase in the temperature of the earth will depend, amongst other things which will be discussed later, on the amount of insolation and the specific heat of the earth. Assuming that the insolation is constant, a surface with a high specific heat warms and cools less quickly than a surface with a low specific heat. For instance, the sea temperature in non-tidal waters hardly changes in the 24 hours. A contributing factor to this small change is its depth. It will be found that, in general, sea temperatures are less than the temperatures of adjacent land by day and greater by night. There is a similar difference in summer and winter. The amount of insolation will depend on the sun's altitude and the degree of cloud cover. It may be noted that the SOLAR CONSTANT, which is the intensity of solar radiation at the outer boundary of the atmosphere, is $1.39 kw/m^2$.

The height of the station under consideration above sea level will be a factor influencing the heating, as will the prevailing wind. Air which has flowed over warm surfaces will have a smaller cooling effect than air which has flowed over cold surfaces.

As the earth is a warm body it radiates heat. This radiation will cool the surface. If the earth is radiating heat at a greater rate than it is receiving it, the net result will be a cooling of the surface. The converse is also true.

In the diagram below the curves of insolation and radiation are shown to increase and decrease at a regular rate for ease of illustration.

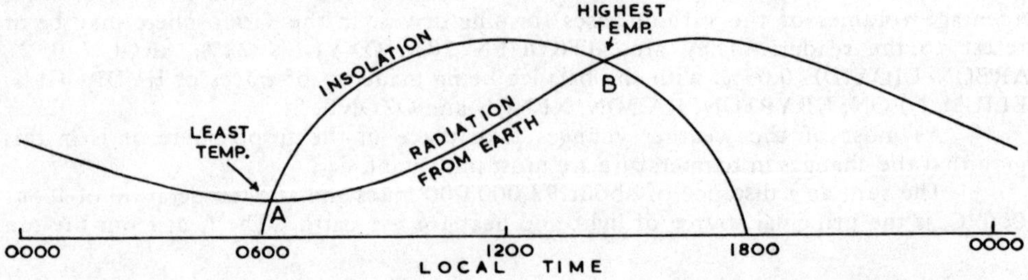

Quite clearly the temperature falls until point A is reached when it begins to rise. The rise continues to point B and then, once again, the temperature falls. It will be noted that the lowest temperature occurs shortly after sunrise whereas the maximum temperature occurs about 1400 hours local time, and not when insolation is minimum and maximum. A similar lag takes place on an annual basis. In northern latitudes the lowest temperatures do not usually occur until February and the maximum temperatures occur in July and early August. The sea temperature lags even further behind the sun. The minimum occuring in March and the maximum in September.

It must be understood that all the above general statements can be considerably modified by local weather conditions, particularly cloud amounts.

The foregoing has described the process by which the earth is heated, but what about the air surrounding it? Heat is transferred to it by one or more of the following means:—

1. Radiation
2. Conduction
3. Convection
4. Turbulence.

1. RADIATION.

As previously mentioned the heat rays from the sun are in short wave form which pass through the atmosphere causing very little heating. Some heating will occur as water vapour in the atmosphere absorbs heat.

The heat rays from the earth are in long wave form, which tend to warm the lower layers of air. If the sky is cloudless most of the earth's radiated heat will go to outer space. However, if there is a cloud cover, most of the heat will be reflected back to earth. The cloud acts as an insulator for the earth.

The heat radiated from the earth on cloudless nights is considerable and a great deal of surface cooling takes place.

2. CONDUCTION.

The heat is transferred from particle to particle. Thus, air which is in contact with a warm surface is warmed; this occurs by day when the sun is shining. Air which is in contact with a cold surface is cooled; this occurs on clear nights. Air masses (see page 54) acquire their characteristics by the same process.

3. CONVECTION.

Air which is warmed expands and consequently its density decreases. This makes it lighter than the unwarmed air surrounding it and the warm air rises. Convectional processes take large amounts of warm air and water vapour from the surface to the upper levels. When the water vapour condenses into water drops and precipitation occurs, the latent heat remains. Most of the atmospheric heating takes place in this way.

Conversely, air which is colder than the surrounding air has a greater density, and as it is heavier, it sinks. An example of this is the Katabatic wind q.v.

4. TURBULENCE.

Air which flows over a rough surface tends to be deflected upwards. The rising air will be replaced by some air from levels up to 600m (2,000 feet) giving an interchange of air between the surface and 600m (2,000 feet). The rising air carries its warmth (acquired by conduction) with it, and the falling air brings its coolness.

LAPSE RATE.

Increase of height above sea level generally means a decrease of temperature. The rate at which the temperature changes with height is known as the lapse rate, an average value being about 0.7°C/100m (3°F/1,000 feet).

ENVIRONMENT LAPSE RATE.

The lapse rate can vary and the Environment Lapse Rate will be referred to when a particular air mass is under consideration. This rate will be dependent on many factors and will vary with the altitude, although it is usually between the S.A.L.R. and D.A.L.R. Plotting temperatures of the air at various heights will give environment curves.

ADIABATIC TEMPERATURE CHANGE.

If a volume of air rises to a region of lower pressure, the volume will increase and, following the gas laws, the temperature of the volume will fall. The converse will occur if the air falls to an area of greater pressure. The change of temperature is solely due to expansion or contraction and no heat has been given to or received from adjacent air. This change of temperature is known as an Adiabatic Temperature Change.

DRY ADIABATIC LAPSE RATE (D.A.L.R.).

When dry air is forced to rise it has been found that it decreases its temperature by 1°C/100m (5.4°F/1,000 feet) and this is known as the Dry Adiabatic Lapse Rate. With dry air which is forced down, an increase at a similar rate is found. Dry air is any air which is not saturated.

SATURATED ADIABATIC LAPSE RATE (S.A.L.R.)

Latent heat is the heat necessary to change 1 kilogram of water to 1 kilogram of vapour at saturation temperature. If a quantity of water changes to vapour an amount of latent heat will have been required. This sort of thing will happen with air at a temperature above dewpoint, (it should be noted that the higher the air temperature the greater is the amount of water vapour that it can hold). Now the latent heat that has been required to change the water to vapour is not lost, as, when the air is cooled to its saturation or dew point temperature, the water vapour condenses into water droplets and the latent heat is released. This is the latent heat of condensation.

When the cooling of the air to its dewpoint temperature is caused by the air rising the latent heat released will modify the overall cooling rate.

For rising air

D.A.L.R.	- 1.0°C/100m
Latent heat	+ 0.5°C/100m (average for near surface in middle latitudes)
S.A.L.R.	- 0.5°C/100m (2.7°F/1,000 feet.)

The S.A.L.R. is a variable quantity depending on the latent heat of the condensing water vapour. The S.A.L.R. is low near the equator and high in polar regions. It will also increase with height.

STABLE AIR.

If air which has been forced to rise (or fall) from its initial level tends to return to that level it is said to be stable. This condition occurs when the environment lapse rate is less than the S.A.L.R.

Figure 1 shows by an unbroken line the environment curve of some tropical air, whose lapse rate is about 0.3°C/100m (2°F/1,000 feet). Suppose some dry air at point A is forced upwards, it will cool at the D.A.L.R. and follow the dotted line. When the rising air reaches the level BB_1 it will be much colder than the environment or surrounding air. As it is colder it is denser, and will want to return to A, a tendency shown by the arrow. If instead of the air rising from A, it fell, then at the level CC_1 the falling air would have warmed adiabatically to a considerably higher temperature than the environment or surrounding air. It would then be less dense and the tendency would be for it to return to A as indicated.

If the air forced to rise or fall from A was saturated instead of dry, the path indicated by the dot/dash line would be followed. Again, for the same reasons as are given above, the tendency is for the air to return to A.

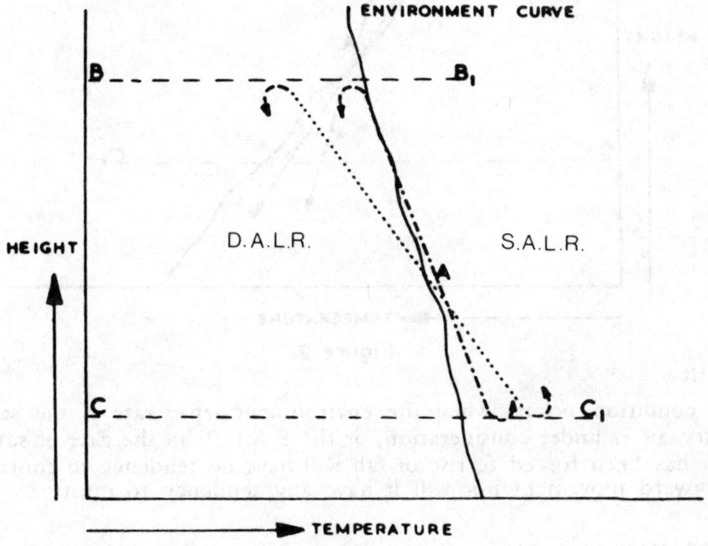

Figure 1.

UNSTABLE AIR.

If air which has been forced to rise (or fall) from its initial level tends to continue its upward (or downward) movement, it is said to be unstable. This condition occurs when the environment lapse rate is greater than the D.A.L.R.

Figure 2 shows by an unbroken line the environment curve of some polar air whose lapse rate is about 1.2°C/100m (6.0°F/1,000 feet).

Suppose some saturated air at point A is forced upward it will cool at the S.A.L.R. and follow the dot/dash line. When this rising air reaches the level BB_1 it will be warmer than the environment or surrounding air. As it is warmer it will be less dense and the tendency is for it to continue upwards. If, instead of the air rising from A, it fell, then at the level CC_1 the falling air would be at a lower temperature than the environment or surrounding air. It would then be denser and it would continue to fall.

If the air forced to rise or fall from A was dry instead of saturated, the path indicated by the dotted line would be followed. Again, for the reasons given above, the rising or falling air continues to rise or fall.

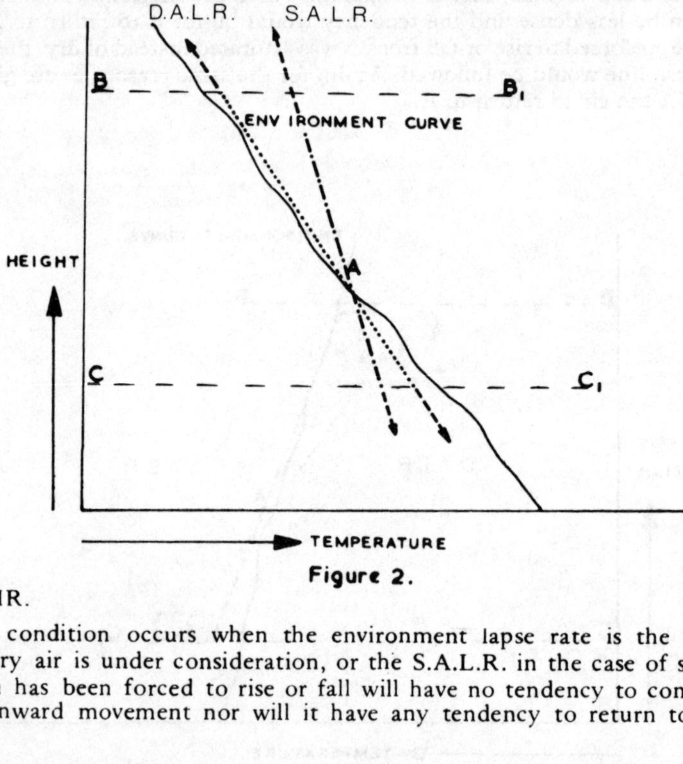

Figure 2.

NEUTRAL AIR.

This condition occurs when the environment lapse rate is the same as the D.A.L.R. if dry air is under consideration, or the S.A.L.R. in the case of saturated air. The air which has been forced to rise or fall will have no tendency to continue its upward or downward movement nor will it have any tendency to return to its original position.

CONDITIONAL INSTABILITY.

This occurs when stable conditions exist for dry air and unstable conditions exist when the air is saturated. Figure 3 illustrates this. The environment curve has a lapse rate between the S.A.L.R. and the D.A.L.R. The dotted line representing dry air forced up or down follows a similar pattern to figure 1, whilst the dot/dash line representing saturated air forced up or down follows a similar pattern to figure 2. This conditionally unstable condition is quite common, the air below the condensation level being stable and that above being unstable.

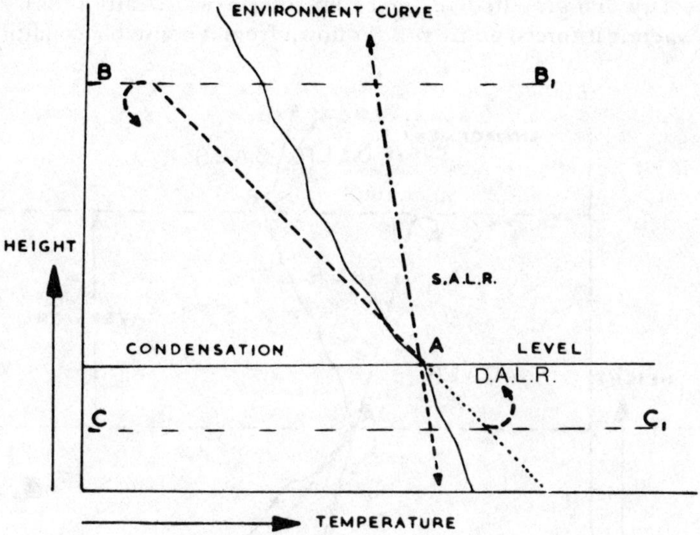

Figure 3.

TEMPERATURE INVERSION

The usual tendency for the temperature to fall with height is sometimes reversed. When the temperature increases with height an inversion is said to exist. An inversion also exists if a layer of air is isothermal.

Inversions may occur at any level at all. They frequently occur at ground level on clear nights. They also occur just above cloud layers.

As an inversion gives rise to very stable conditions, convection and turbulence are damped out and there will be no upward movement of air through the inversion.

Inversions may be caused in anticyclonic conditions by subsiding air warming adiabatically to a temperature above that of the lower layers of air.

Figure 4 shows a curve for the environment where there is an inversion above the surface. Dry and saturated air forced up from A will clearly return showing stable conditions, whereas if forced up from B (or down from A) unstable conditions exist.

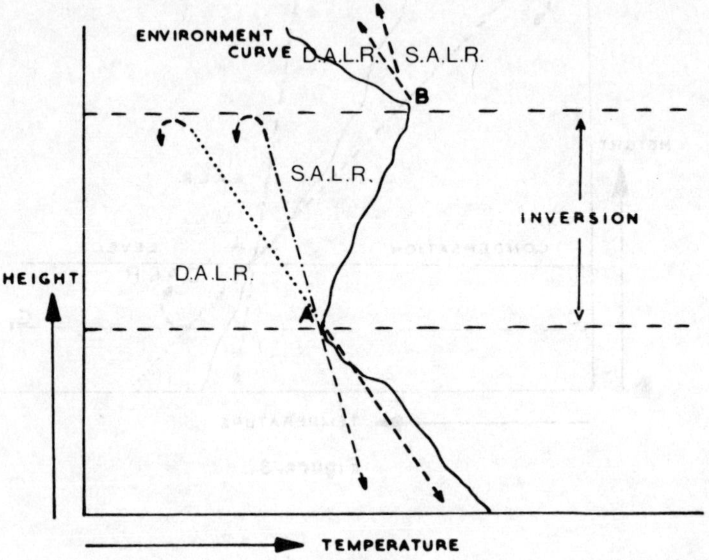

Figure 4.

TEPHIGRAM.

For accurate forecasting a knowledge of upper air conditions is essential. If upper air temperatures are known (from Radio-sonde measurements) the other conditions can be deduced from a tephigram.

The tephigram is a graticule formed by plotting a series of straight lines representing temperature, water vapour content, and dry adiabatics (also called lines of constant potential temperature) against curves (almost horizontal straight lines) which indicate height-pressure for a standard condition of the atmosphere (known as the ICAN atmosphere). Also included on the graticule are curves to show saturated adiabatics for the same standard atmosphere. Such curves are shown on Metform 2810 but they are unlikely to be found aboard ship.

III
Cloud and Precipitation

When air is cooled below its dewpoint the water vapour therein starts to condense out into water droplets. The water droplets form either fog or cloud depending on the process by which the air is cooled. It is generally understood that the fog is formed when the cooling of the air takes place at the surface by conductive processes, whereas cloud generally forms above the surface due to the adiabatic cooling of the rising air. The cooling of the air occurs as the air is forced to rise as follows :–

(a) at a warm front
(b) at a cold front
(c) by convectional processes
(d) by high ground
(e) by turbulence

Contributory causes in cloud formation may be :–

(i) radiation of heat by water vapour in the atmosphere
(ii) mixing of two masses of nearly saturated air at different temperatures.

Clouds are classed as being either high above 6000 metres (20,000 feet), medium 2500 - 6000 metres (8,000 - 20,000 feet) or low below 2500 metres (8,000 feet).

There are several types in each class, the main ones being shown in the following table :–

CLASS	STRATIFORM (Stable air conditions)	FREE CONVECTIVE (Unstable air conditions)	LIMITED CONVECTIVE (Air conditions tending to be unstable)
LOW	Stratus (St)	Cumulus (Cu)	Stratocumulus (Sc)
	Nimbostratus (Ns)	Cumulonimbus (Cb)	
MEDIUM	Altostratus (As)		Altocumulus (Ac)
HIGH	Cirrostratus (Cs)	Cirrus (Ci)	Cirrocumulus (Cc)

There are other forms of cloud such as :–

Fractostratus or Scud. Fracto means broken and may be applied to other types of clouds besides stratus.

Lenticularis. Indicates clouds of lens or airship shape.

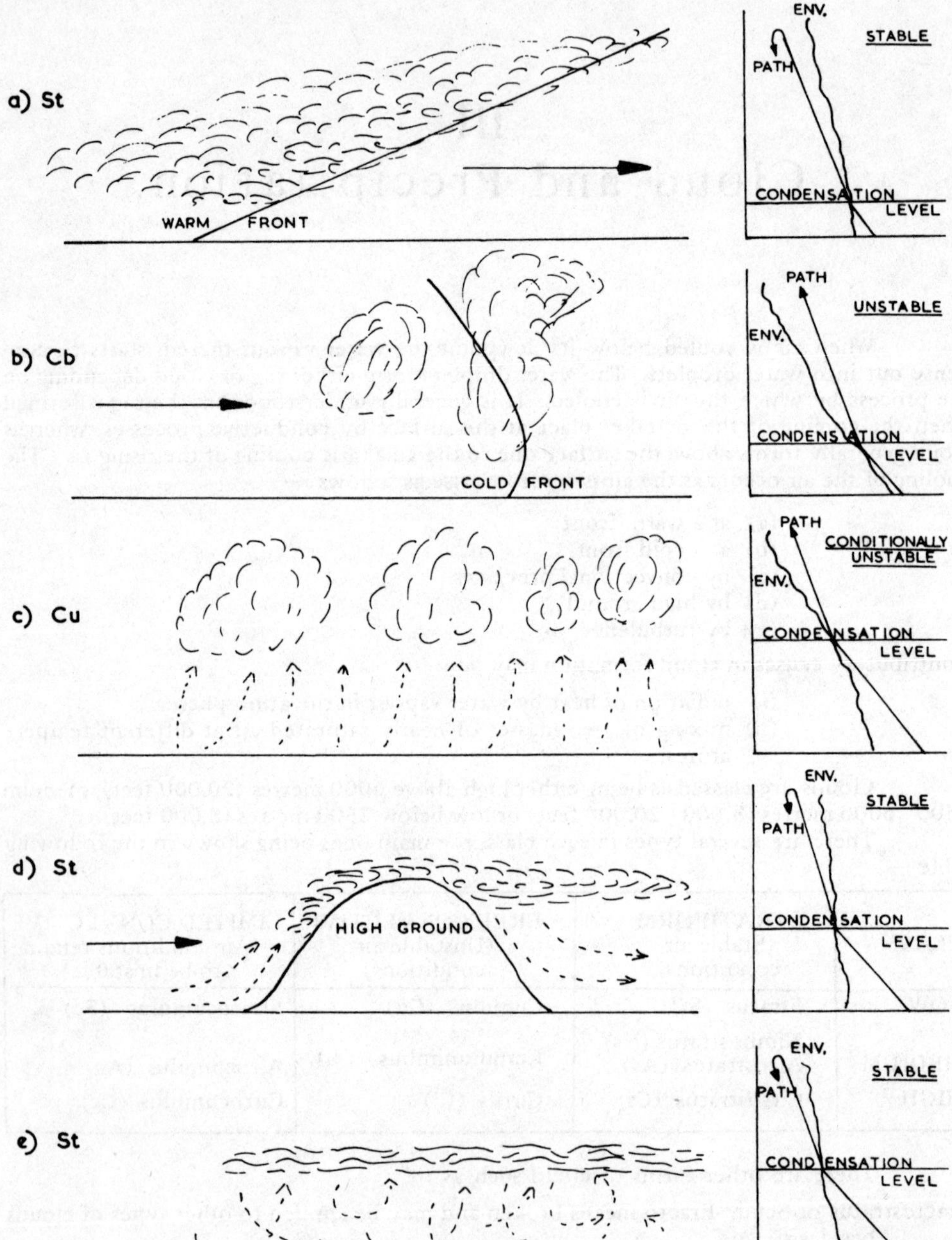
a) St
WARM FRONT
ENV.
STABLE
PATH
CONDENSATION LEVEL
b) Cb
COLD FRONT
PATH
UNSTABLE
ENV.
CONDENSATION LEVEL
c) Cu
PATH
CONDITIONALLY UNSTABLE
ENV.
CONDENSATION LEVEL
d) St
HIGH GROUND
ENV.
STABLE
PATH
CONDENSATION LEVEL
e) St
ENV.
STABLE
PATH
CONDENSATION LEVEL

Castellatus. Here the tops of the cloud may extend vertically to form 'castles'. This often occurs with altocumulus.

Mammatus. The lower side of the cloud projects to form festoons. Not very common, but may be found with unstable conditions, where there are strong down as well as up currents.

The water vapour mentioned earlier cannot condense into large size water drops unless condensation nuclei are present. These nuclei, which may be salt or dust particles, are always present although the numbers will vary from about 1000 per cm^3 over the sea to about 150,000 per cm^3 over industrial areas.

Water drops of less than 0.1mm diameter cannot reach the ground as they evaporate on the way. These drops are cloud drops and their average diameter is 0.01mm.

Drops over 0.1mm are drizzle drops and over 0.5mm are rain drops. The maximum size to which a drop can grow is 5.5mm. The process of growth of these water drops has brought forth many theories one of which is that in the original stages of condensation taking place, some drops form larger than the rest. Their growth, due to the attraction of the smaller drops, is then assured.

Another theory is that for growth to take place, part of the cloud must be above the freezing level, in which case the ice crystal will grow at the expense of water drops. When these get sufficiently large they will fall through the cloud, warming and melting as they fall, until they arrive at the lower level as rain. This is known as the BERGERON PROCESS. It should be noted that the freezing level over the British Isles is about 1000m (3,000 ft.) in winter and about 2500m (8,000 ft.) in summer. This latter theory would account for the tendency to greater rainfall in winter than in summer.

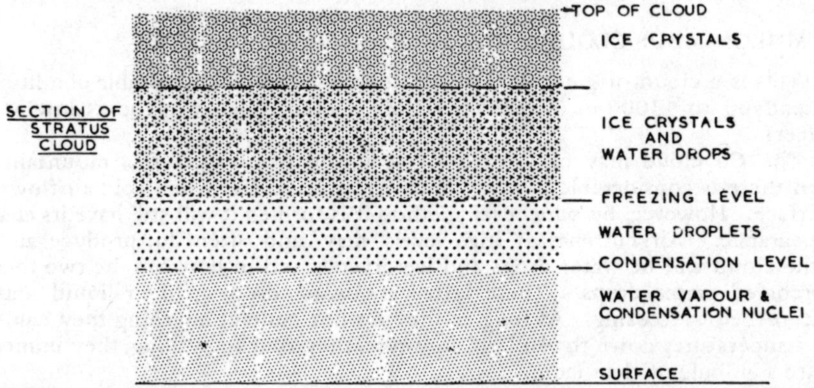

Rain is usually described as being either Frontal, Convectional or Orographic. This is in accordance with the process of cloud formation(see page 23). Appreciable rainfall is unlikely with turbulent cloud. FRONTAL RAIN occurs at a warm front. The precipitation commences from Altostratus cloud and increases in intensity as the lower nimbostratus cloud comes in.

Where the front is not so marked OCCASIONAL PRECIPITATION may occur. This has the same characteristics as frontal rain and lasts less than 30 minutes.

CONVECTIONAL RAIN falls when the strong up currents, within the cumulus type cloud, cease or are so reduced that they can no longer keep the rain drops within the cloud. It may be noted that an up current of 8m/s (25 feet per second) will keep the maximum sized raindrops from falling. If this velocity is slightly reduced, only the largest sized drops will fall and the smaller drops will remain in the cloud. A feature of convectional type rain is that rainfall always starts heavily and often with little warning. Another name for convectional rain is showers. These last less than 30 minutes although showers frequently merge to give a period of prolonged rainfall.

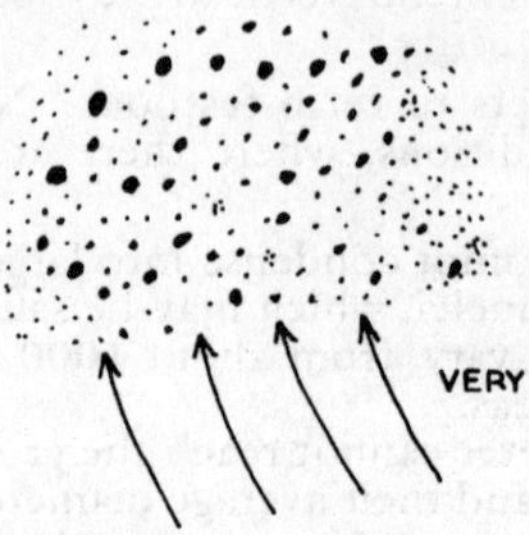

OROGRAPHIC RAIN. This falls on the weather side of mountain ranges and gives some of the heaviest rain known.

SNOW. When water vapour condenses at a temperature below freezing point ice crystals form. These join together to form snowflakes. If the air temperature is much above freezing point the flakes will melt. In general the temperature near the surface must not exceed 3°C. (37°F.) for snow to fall. It should be understood that at temperatures less than 3°C rain is still very likely and is not necessarily replaced by snow.

SLEET. This is partially melted snow.

THE CUMULONIMBUS CLOUD.

This is a cloud of great vertical development formed in unstable conditions and may extend up to 13000 m (40,000 ft.) from a base which may be as low as 500m (1,500 feet).

The Cb cloud may be formed (i) at a cold front, (ii) at a mountain range, (iii) when there is considerable heating of the ground, or (iv) when cold air flows over a warm surface. However, by whichever method it is formed it will still have its characteristic appearance. Air currents at high levels may, and often do, produce an 'anvil'. Within the cloud will be water drops and ice crystals and in between the two there may be supercooled water drops. These latter are water drops in their liquid state at a temperature below freezing. As long as they do not contact anything they can remain liquid at temperatures down to -40°. If, however, they touch anything, they immediately freeze into a globule of clear ice.

HAIL. Due to the turbulence within the Cb cloud ice crystals frequently contact the supercooled water drops. When this occurs the supercooled drop freezes to give a layer of clear ice round the ice crystal. This is the start of the hailstone, which will then grow as convection currents carry it through the cloud.

During this process the stone acquires alternate coats, one of clear ice (from the supercooled drops) and one opaque ice (from the ice crystals). If a hailstone is cut open the various layers can be seen, and the number of times it has been carried up and down can be estimated.

When the hailstone can no longer be supported within the cloud it will fall to the ground. The hailstone varies considerably in size being anything from 5mm to 50mm in diameter.

THUNDERSTORM. When raindrops reach their maximum size of 5.5mm they break up. This splitting gives the air a negative electrical charge whilst the raindrop retains a positive electrical charge.

This action takes place within the Cb cloud and it is found that there is a large positive charge near the top of the cloud. At the bottom there is a large negative charge with a small positive charge due to the raindrops.

When there is sufficient difference of potential between the top and bottom of the cloud, or between two clouds, or between a cloud and the earth, there is a visible electrical discharge which is called LIGHTNING.

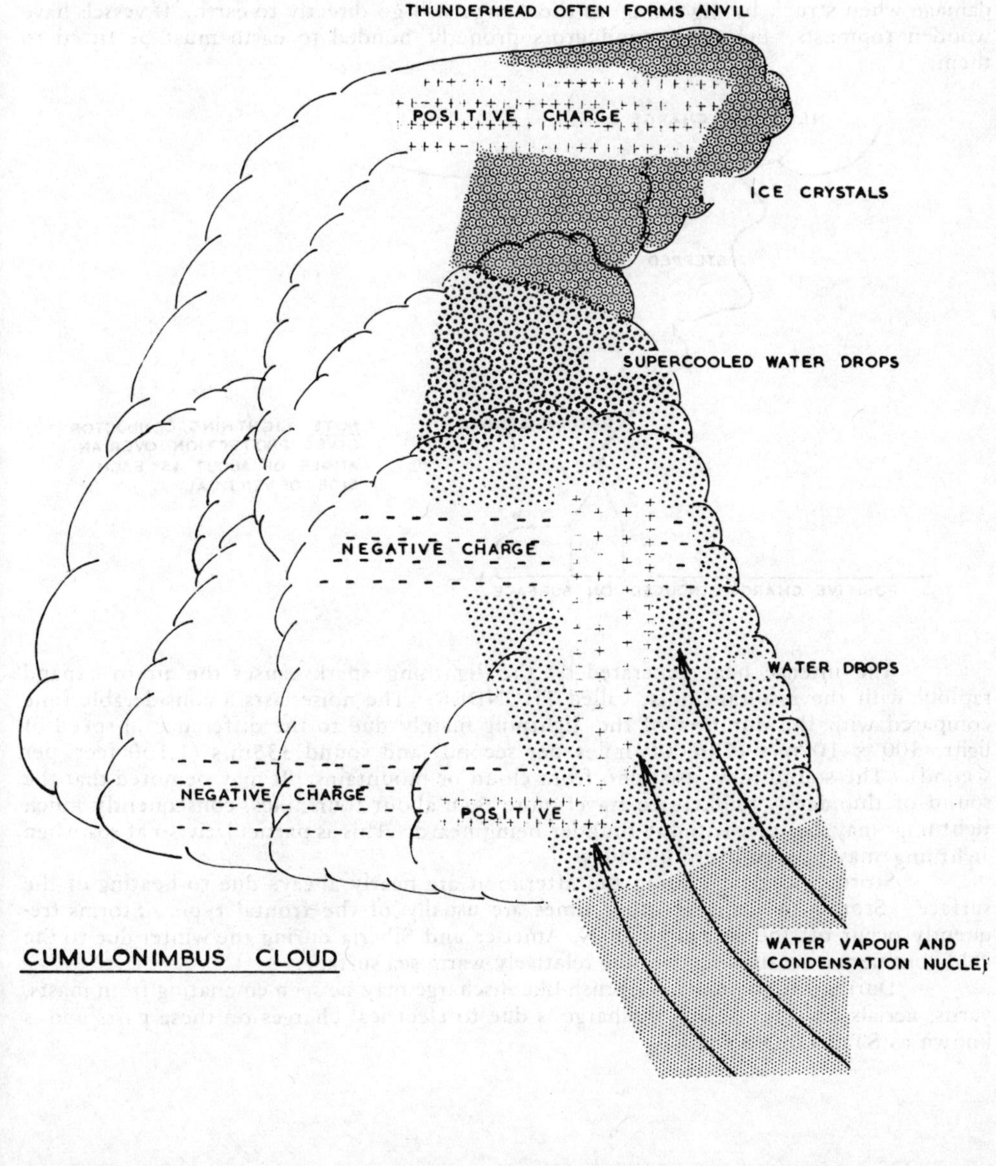

CUMULONIMBUS CLOUD

When the P.D. is between the cloud and the earth, the negative charge on the underside of the cloud induces a positive charge on the earth. A leader stroke known as a stepped leader finds a path from the cloud to a point near the earth when an upward stroke from the earth comes up to meet it. This up stroke continues along the ionised path made by the stepped leader. There will probably be many up and down strokes in the space of a microsecond or so, each striking the same place. The lightning conductor concentrates the charge and ensures that the lightning will be discharged without damage to the building. In any place the upward stroke is likely to originate from the highest point in the vicinity of the downward leader. Ships will not suffer any damage when struck by lightning as the charge will go directly to earth. If vessels have wooden topmasts, lightning conductors properly bonded to earth must be fitted to them.

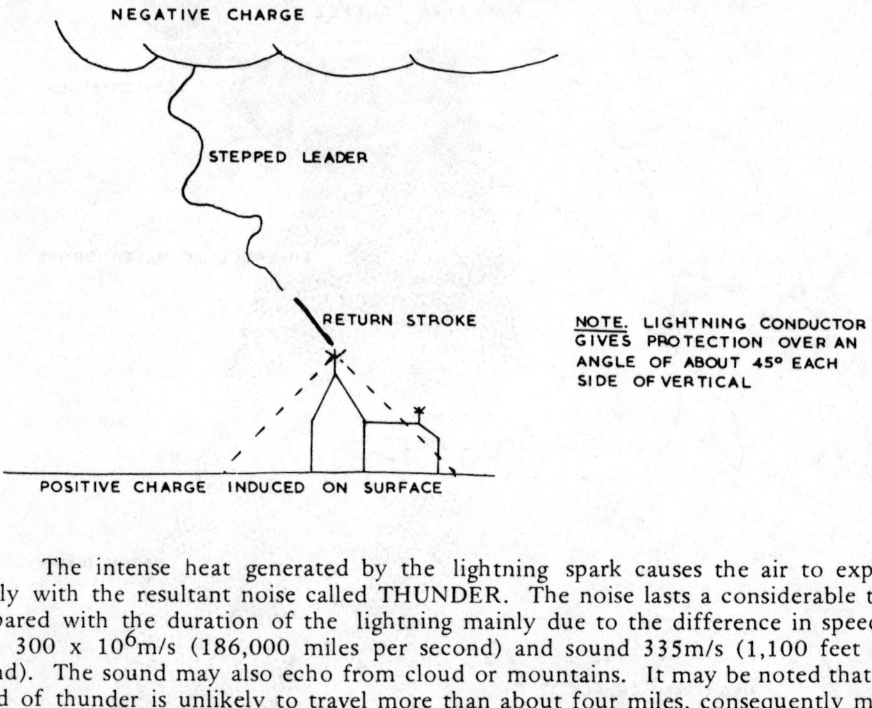

The intense heat generated by the lightning spark causes the air to expand rapidly with the resultant noise called THUNDER. The noise lasts a considerable time compared with the duration of the lightning mainly due to the difference in speed of light, 300×10^6m/s (186,000 miles per second) and sound 335m/s (1,100 feet per second). The sound may also echo from cloud or mountains. It may be noted that the sound of thunder is unlikely to travel more than about four miles, consequently much lightning may be seen without thunder being heard. This is particularly so at sea when lightning may be seen over 100 miles.

Storms occuring during the afternoon are nearly always due to heating of the surface. Storms occuring at other times are usually of the frontal type. Storms frequently occur off the east coast of N. America and Siberia during the winter due to the cold continental air flowing over the relatively warm sea surface.

During thunderstorms a brush-like discharge may be seen emanating from masts, yards, aerials or stays. This discharge is due to electrical charges on these parts and is known as ST. ELMO'S FIRE.

So far the phenomena caused by air cooling through rising have been considered. If air is cooled below its dewpoint by conduction (i.e. contact with a cold surface) then the water vapour will condense into droplets on or near that surface. As has been seen on page 17, the radiation from the earth on cloudless nights will cause considerable surface cooling, with the possible deposit of water droplets if the air near the earth is cooled until saturated.

DEW. The water droplets deposited on the surface by the above process. Clear skies and practically no wind are necessary conditions for its formation.

HOAR FROST. Formation is the same as for dew except that the dewpoint temperature is below freezing. Hoar frost may also be formed by the freezing of water droplets which were deposited as dew.

FOG occurs when visibility is reduced to less than 1000 metres. There are different means by which fog is formed and these are detailed below.

MIST is similar to fog but here the visibility is 1000m to 2000m. It should be noted that HAZE giving the same visibility as mist is formed of dust particles not water droplets.

RADIATION FOG occurs over land at night when radiation from the earth has cooled its surface to below the dewpoint of the air adjacent to it. Conduction cools this air until condensation takes place. Slight turbulence (wind 3–7 knots) will carry this cold saturated air to higher levels and will bring to the surface drier air, which will in turn be cooled and saturated. As the process continues the whole lower layer of air will become saturated. A temperature inversion is often associated with radiation fog, which, acting as a "lid", keeps the fog low down and causes it to thicken.

Stronger winds will disperse the fog as will the heat from the sun, provided the fog layer is not too thick. In winter there is rarely sufficient heat to disperse fog and persistent fogs occur. If over industrial areas, "SMOG" frequently forms as the smoke from chimneys is trapped under the temperature inversion.

This type of fog may occur overnight on the banks of rivers but usually disperses fairly quickly in the morning.

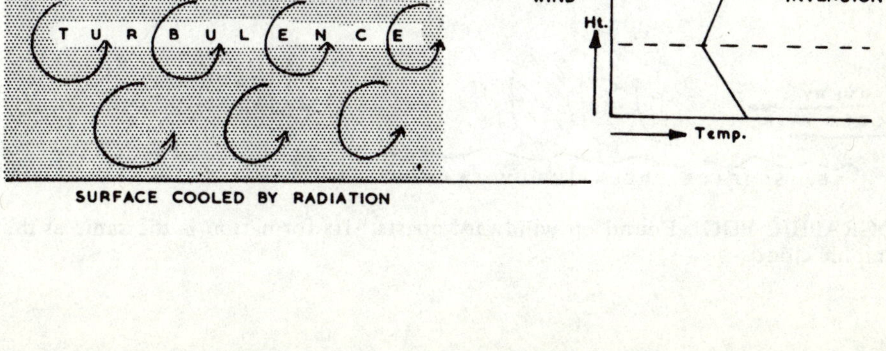

ADVECTION FOG. This occurs over the sea when warm tropical maritime air is cooled below its dewpoint by conduction from a cold sea surface. The likelihood of fog occuring may be judged if graphs of the air dewpoint temperatures and the sea temperatures are drawn and plotted against time. In the graph below fog is likely to occur about 2000 hours, if the present trends are maintained. A change of course to the southward would alter conditions considerably. WINDS 15-17 KN

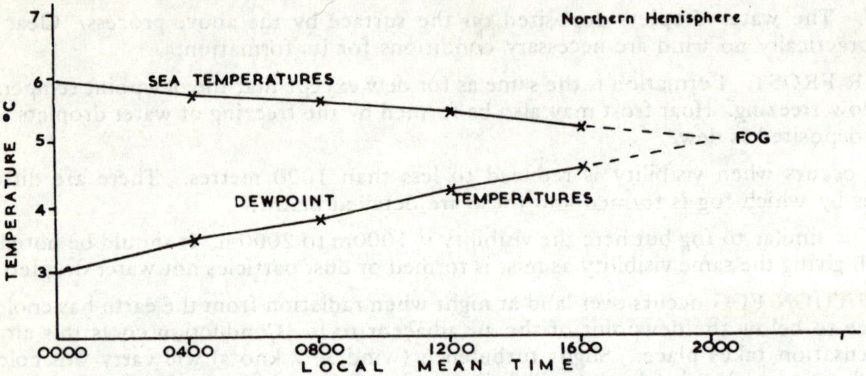

In general terms fog is likely to occur if tropical maritime air crosses seven sea isotherms from its source

Sea fog occurs mainly in the summer months in the following areas :— British Isles, Banks of Newfoundland, West Coast of North America, Japan, West Coasts of South Africa and South America.

If reference is made to the current chart on pages 44 and 45, it will be seen that the areas mentioned have cold sea currents flowing through them.

SEA SMOKE OR ARCTIC SMOKE. This occurs over the sea when the sea is warm compared with the air which flows over it. The air next to the water is warmed by conduction. This warmed air takes up water vapour before rising through the colder air above, where cooling condenses out the water vapour. The overall effect is that of visible steaming.

This frequently occurs in winter over rivers and also off the East Coasts of continents.

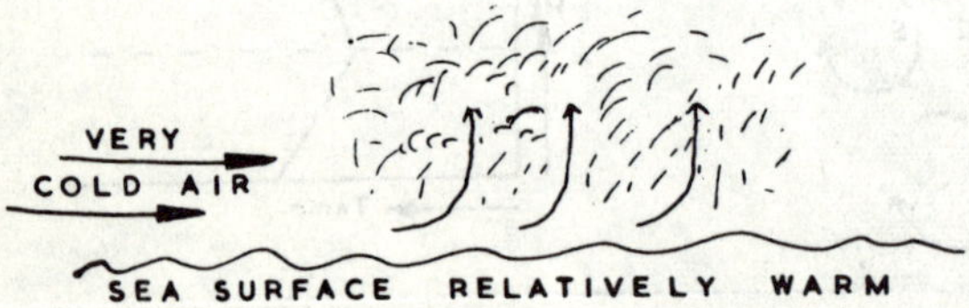

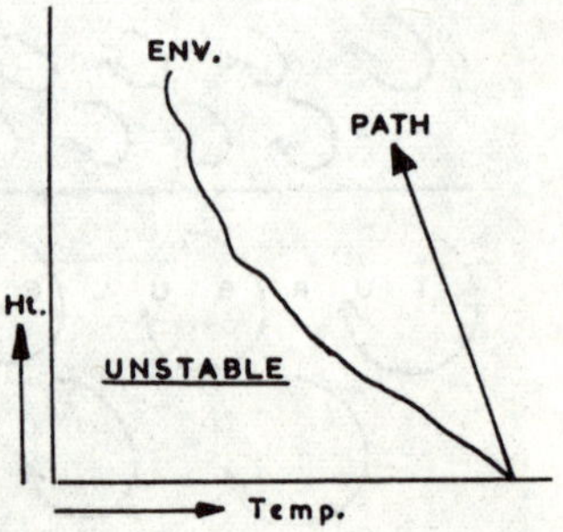

OROGRAPHIC FOG. Found on windward coasts. Its formation is the same as that of orographic cloud.

Other phenomena associated with water drops are :—

RIME. This occurs when supercooled water drops (often in fog) freeze into opaque ice on coming into contact with rigging on ships or trees or telegraph wires on land. The ice formed grows out to windward.

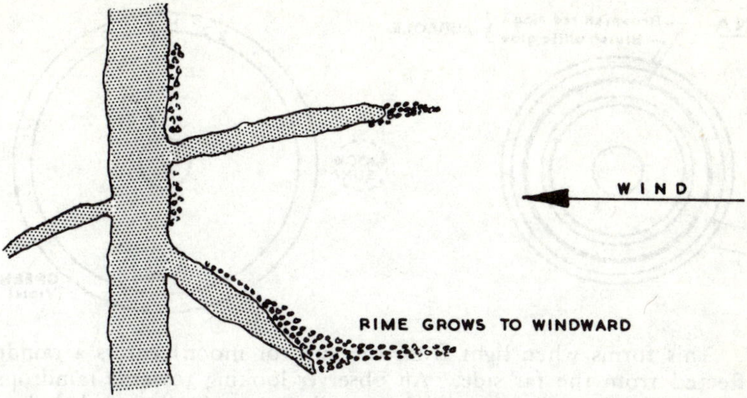

GLAZED FROST OR BLACK FROST. This occurs when water drops fall onto a surface which is below freezing. A layer of clear ice is formed and this creates chaos on roads as the ice conditions cannot be seen. However, the effect at sea is much more serious as, if water falls onto decks, rigging and superstructures which are below freezing, the weight of the ice which forms can affect the stability so that the safety of the vessel is endangered. Although ice axes and steam hoses are partially effective in removing the unwanted ice, the only sure way of combating black frost is to steam into warmer conditions.

OPTICAL PHENOMENA of various sorts are associated with water vapour or ice crystals in the atmosphere, others being associated with dust particles. Some of the more common ones are described briefly below.

AUREOLE. A brownish red ring surrounding a bluish-white glow 2°—3° in diameter around the sun or moon. The light is diffracted by water drops.

BISHOP'S RING. A reddish brown ring 10°—20° radius, occasionally seen round the sun with a clear sky. The light is defracted by dust particles.

CORONA. A number of coloured rings round the sun or moon which appear outside the aureole. The radius is generally less than 10°, but this differs with the size of the water drops from which the light is diffracted. The smaller the drops the larger is the radius of the corona. If colours are visible, violet will be nearest the sun and red away from it. Colour is rarely visible with lunar coronae.

HALO. Occurs when light from the sun or moon is refracted through ice crystals. In its most common form it is a ring of 22° radius. When the colours can be distinguished red is nearest to the sun and violet on the side of the halo furthest from the sun. A halo of 46° radius is sometimes seen. The difference in radii is due to the different angles which the faces of the ice crystal make with one another.

MOCK SUN OR PARHELION. This may form at the same altitude as the sun about 22°—36° from the real sun. When coloured, red (as with the halo) is nearest the sun. Mock moons or paraselenae may also occur. The formation is by refraction of light through ice crystals.

PARHELIC CIRCLE OR MOCK SUN RING. This is another of the halo phenomena and is a white circle parallel to the horizon at the same altitude as the sun. Its width is about 1° and, under ideal conditions, can be seen all round the horizon.

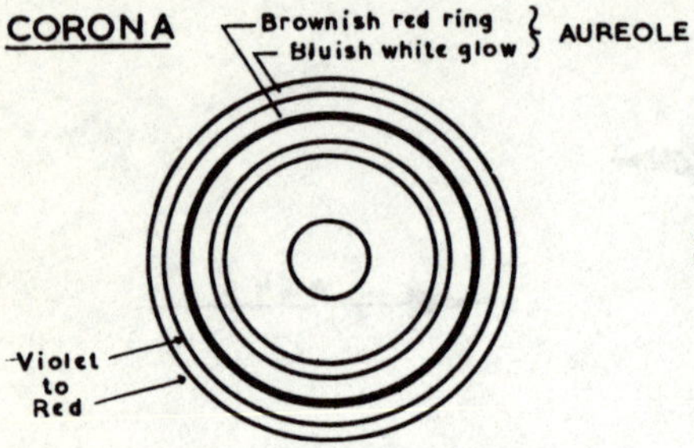

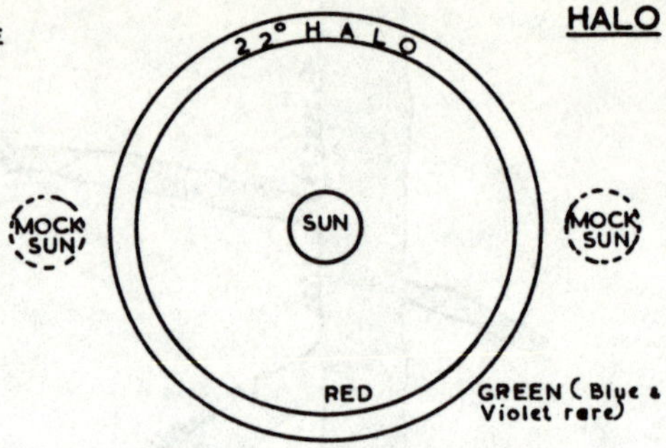

RAINBOW. This forms when light from the sun (or moon) enters a raindrop and is reflected from the far side. An observer looking towards raindrops with his back to the sun will see a rainbow radius about 42° provided that the sun's altitude does not exceed 42°. The colours of a primary bow are red, orange, yellow, green, blue, indigo and violet, the red being on the outside. The secondary bow may form about 9° outside the primary. The colours of the secondary bow are in the opposite order to those of the primary. A secondary bow cannot form if the sun's altitude is more than 54°. The colours seen depend on the intensity of light and if the rainbow is formed by moonlight it will nearly always appear to be white.

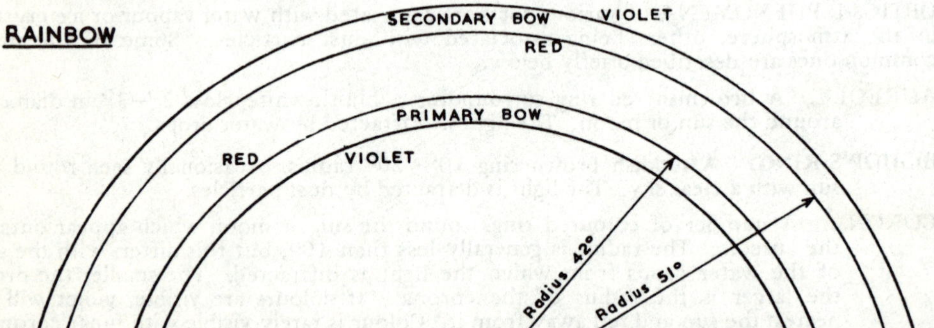

A well known phenomena which is unconnected with either waterdrops or dust particles is the MIRAGE. This occurs when there is abnormal refraction. This may occur with a temperature inversion, particularly in the polar regions. With these conditions a distant ship may be seen in triplicate with one of the images inverted. Mirages formed when the air next to the ground is less dense than that above appear to cut off the bottom of objects so that land and ships appear to float in the air.

IV
Wind

Consider the two columns of air AB and CD in the figure below. These are of equal cross sectional area, height and temperature, so that the pressures at A and C are equal, as are those at B and D. Suppose the column AB is warmed it will expand to AE, if the area remains constant, and although the pressure at A will remain unaltered, the pressure at B will have increased relative to that at D, and to equalise these air will flow from B to D. This transfer of air from column AB to CD will cause a drop in pressure at A and a rise at C, and air will now flow from C to A.

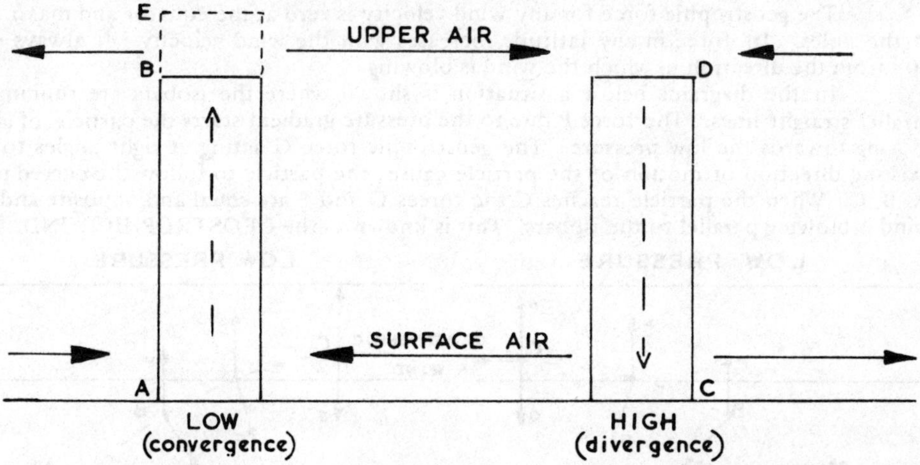

This illustrates the vertical circulation of air in that the outflowing or diverging air from C is replaced by subsidising air from above, whilst the inflowing or converging air at A escapes to the upper levels and there completes the circulation back to D.

The rate of flow of air from C to A will depend on the pressure gradient (see page 3) between C and A.

The movement of air straight from a high pressure to a low pressure only occurs either very close to the equator or in places where the high and low pressure areas are small and their centres are near to each other, (see land and sea breezes). Elsewhere, due to the rotation of the earth, there is a deflective force which prevents the air flowing directly from high pressure to low. This force is known as the GEOSTROPHIC or CORIOLIS force.

Figure (i) below is a view of the earth from a point above the North Pole (Pn) whilst figure (ii) is the earth viewed from a point above the South Pole (Ps). In each case a parcel of air is sent from point A towards a point C in space. At the outset point B on the earth is in transit with A and C, but by the time the parcel of air reaches a point above B, B will have moved to B_1 due to the earth's rotation. To an observer at B in the Northern Hemisphere the air appears to have been deflected to the right, whereas in the Southern Hemisphere it appears to have been deflected to the left.

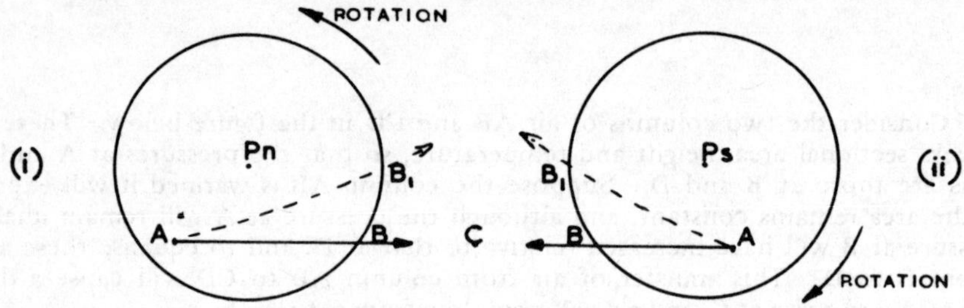

The geostrophic force for any wind velocity is zero at the equator and maximum at the poles. Its force in any latitude increases with the wind velocity. It always acts 90° from the direction in which the wind is blowing.

In the diagrams below a situation is shown where the isobars are running in parallel straight lines. The force P due to the pressure gradient starts the particle of air A moving towards the low pressure. The geostrophic force G acting at right angles to the existing direction of motion of the particle causes the particle to follow the curved path A, B, C. When the particle reaches C the forces G and P are equal and opposite and the wind is blowing parallel to the isobars. This is known as the GEOSTROPHIC WIND.

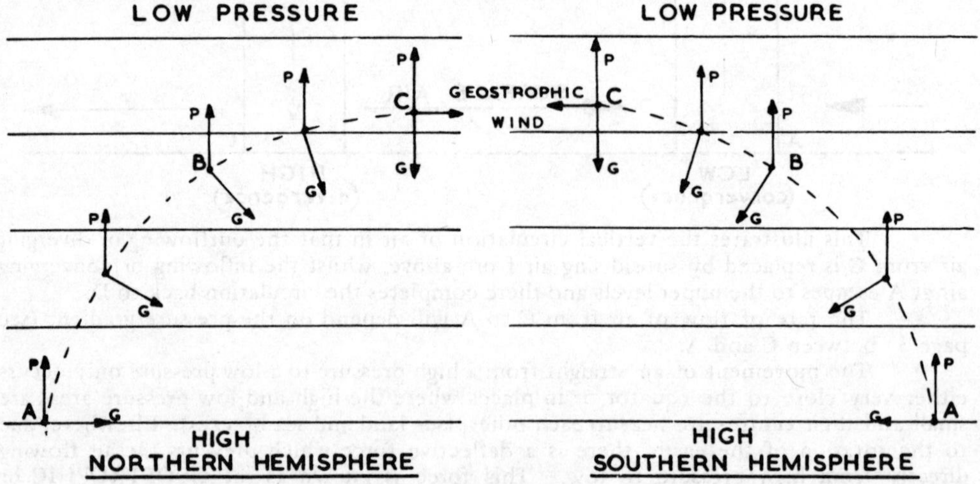

It can be seen from the diagram that if an observer faces the wind in the Northern Hemisphere, the low pressure lies to his right, whilst in the Southern hemisphere the low pressure lies to his left. These facts were first propounded by Buys Ballot and are known as BUYS BALLOT'S LAW.

If the pressure gradient is known the geostrophic wind speed can be calculated from the formula :—

$$V_g = \frac{P}{2\rho\,\omega \sin\phi}$$

Where V_g = the geostrophic wind speed
P = the pressure gradient
ρ = the density of air
$\omega\sin\phi$ = rate of turntable motion in latitude ϕ

If a scale distance is to be used for isobar spacing a diagram showing wind speeds for different pressure gradients can be drawn. Such a diagram is shown on Metform 1258 – the North Atlantic Plotting Chart – and is reproduced below with the permission of the controller of H.M. Stationery Office and the Director of the Meteorological Office.

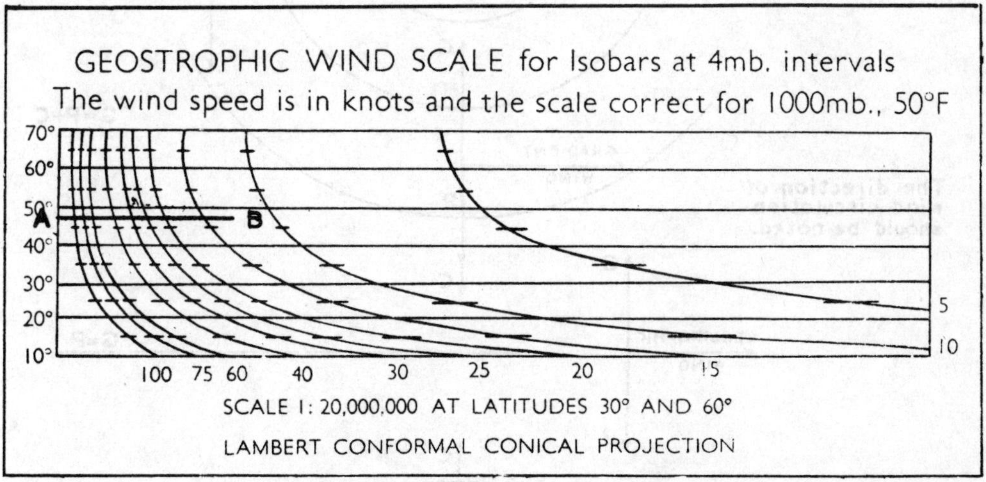

To use the scale take a pair of dividers and measure the perpendicular distance between isobars spaced at 4 millibar intervals at the required position on the plotting chart (i.e. after drawing the synoptic chart as detailed in Chapter VIII). The distance is now transferred to the geostrophic wind scale so that one leg of the dividers is placed on the vertical scale on the latitude of the position, whilst the other will be at a point on or between the wind speed curves horizontally to the right of the vertical. Follow the curve down to the base line where the wind speed can be read off either directly or by interpolation.

For example if the distance between isobars 4mb apart in latitude 48°N is represented by AB, the geostrophic wind is 13 knots.

It must be emphasised that this is the geostrophic wind which blows at and above 600m (2000 ft.), the surface wind speed is about 2/3 that of the geostrophic.

It is useful to know that for equal pressure gradients the wind velocity in the tropics (lat.15°) is three times as great as that in temperate latitudes (lat.52°).

If the isobars are curved, as around a high or low pressure area, a further force will come into being. This force, an outward force from the centre of high or low pressure, is called the CYCLOSTROPHIC FORCE.

In the diagrams below it can be seen that the cyclostrophic force C acts in the same direction as P with the high pressure circulation, so that G = P + C, and in the opposite direction to P with the low pressure system such that G = P - C. The resulting wind is the GRADIENT WIND. It may be noted that the wind will be stronger around the high pressure system than around the low pressure system always provided that the pressure gradient is the same in each case.

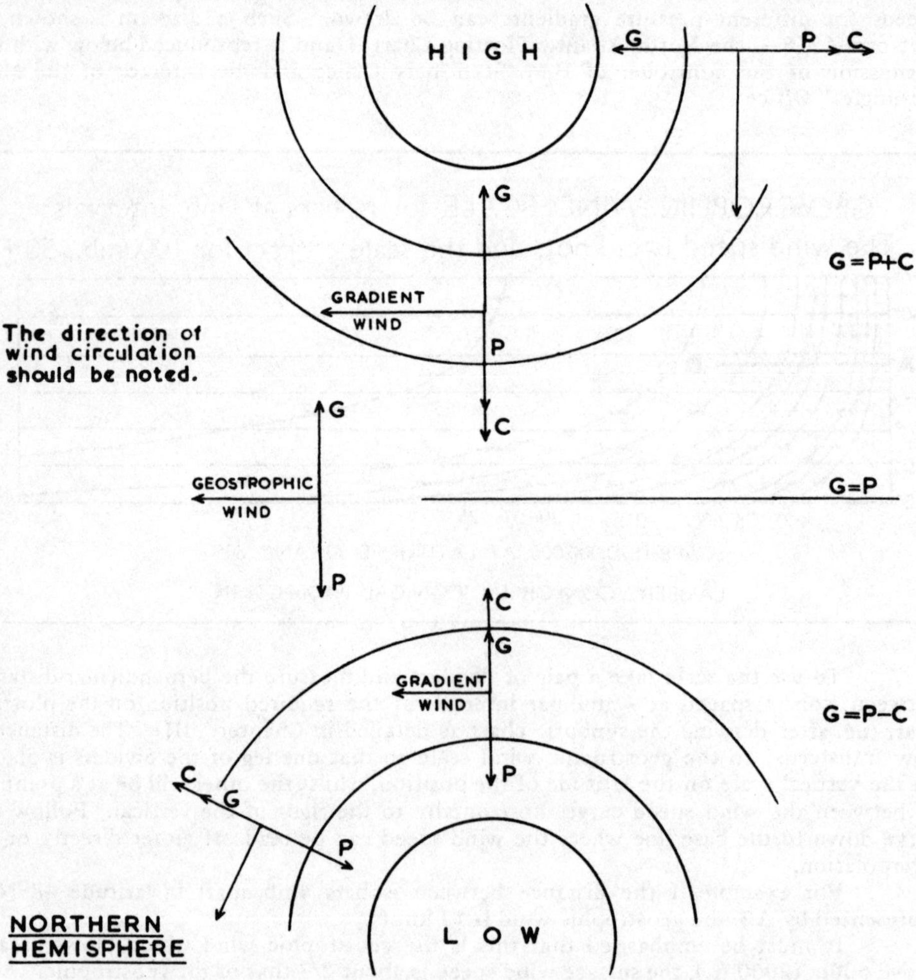

The winds, so far referred to, have been unaffected by friction; such winds will blow above 600 metres(2,000 ft.). The effect of friction is to reduce the wind velocity. The reduction gets progressively greater between 600 metres (2,000 ft.) and sea level. At sea level the wind velocity is only half to two thirds that at 600m (2,000 ft.) depending on the type of surface, the reduction being greater over the land than over the sea.

The sketch shows that although the wind velocity is reduced by friction the pressure gradient is unchanged. In order that the equilibrium of point A shall be maintained, the direction of the wind must be moved towards the force P. Thus the surface wind is inclined towards the low pressure and away from the high pressure. The angle which the wind makes with the isobars is known as the ANGLE OF INDRAUGHT.

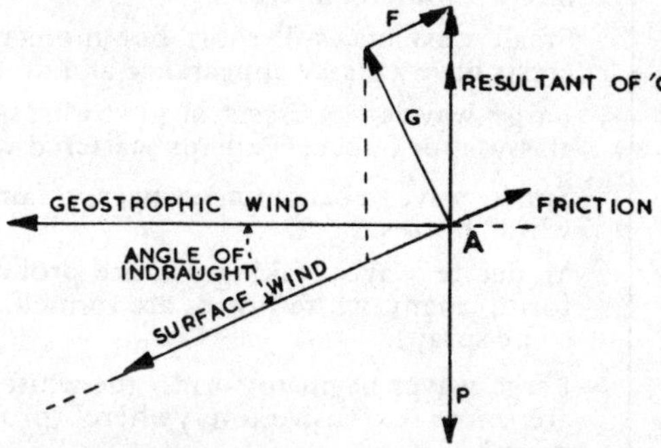

During the daytime turbulence due to surface heating causes the higher velocity winds at 600 metres (2,000 ft.) to be brought to the surface and it will be noted that during this time the surface wind veers and increases in velocity. By night turbulence is damped out and it will be noted that the surface wind backs and decreases in velocity. The effects mentioned above are hardly noticeable over the sea as turbulence due to surface heating is negligible.

The direction of the true wind should be noted in the logbook and in weather messages. It may be estimated by noting the direction of the waves. The approximate speed of the wind may be estimated from the appearance of the sea (see Beaufort scale). The apparent wind direction and speed is the resultant of the ship's course and speed and the true wind direction and speed. If any two of the foregoing are known the third can be found by plotting a triangle of velocities.

E.g. An officer on a vessel steaming NNE at 11½ knots observes the apparent wind to be 6 points on the port bow at 15 knots. What would be the true wind direction and speed?

AB represents the ship's course and speed. This is the wind caused by the ship; hence the direction of the arrow head is opposite to the course. AC represents the apparent wind direction and speed. A double arrow head denotes this as it is the resultant of the wind caused by the vessel and the true wind. Joining B to C completes the triangle of velocities. BC will be the direction and speed of the true wind.

True Wind West 15 knots.

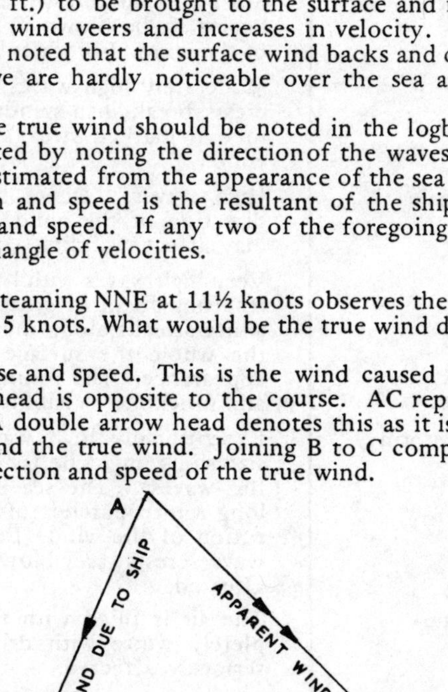

BEAUFORT WIND SCALE

Beaufort Scale Number	Description and limit of wind speed in knots.	Sea Criterion
0	Calm Less than 1	Sea like a mirror.
1	Light air 1–3	Ripples with the appearance of scales are formed but without foam crests.
2	Light breeze 4–6	Small wavelets, still short but more pronounced, crests have a glassy appearance and do not break.
3	Gentle breeze 7–10	Large wavelets. Crests begin to break. Foam of glassy appearance. Perhaps scattered white horses.
4	Moderate breeze 11–16	Small waves, becoming longer; fairly frequent white horses.
5	Fresh breeze 17–21	Moderate waves, taking a more pronounced long form; many white horses are formed. (Chance of some spray).
6	Strong breeze 22–27	Large waves begin to form; the white foam crests are more extensive everywhere (probably some spray).
7	Near gale 28–33	Sea heaps up and white foam from breaking waves begins to be blown in streaks along the direction of the wind. (spindrift begins to be seen).
8	Gale 34–40	Moderately high waves of greater length; edges of crests break into spindrift. The foam is blown in well marked streaks along the direction of the wind.
9	Strong gale 41–47	High waves. Dense streaks of foam along the direction of the wind. Sea begins to roll. Spray may affect visibility.
10	Storm 48–55	Very high waves with long overhanging crests. The resulting foam in great patches is blown in dense white streaks along the direction of the wind. On the whole the surface of the sea takes a white appearance. The rolling of the sea becomes heavy and shocklike. Visibility affected.
11	Violent storm 56–63	Exceptionally high waves. (Small and medium-size ships might be for a time lost to view behind the waves). The sea is completely covered with long white patches of foam lying along the direction of the wind. Everywhere the edges of the wave crests are blown into froth. Visibility affected.
12	Hurricane 64–71	The air is filled with foam and spray. Sea completely white with driving spray; visibility very seriously affected.

There is an extended scale for exceptional wind as follows :—

Force	Knots
13	72 – 80
14	81 – 89
15	90 – 99
16	100 – 109
17	110 – 118

STORM WARNINGS are hoisted at stations on the coasts of the British Isles when winds of force 8 or more are expected shortly within 50 – 100 miles of the station. The day signal consists of a cone 1 metre (3 feet) wide and 1 metre (3 feet) high. A night time signal is hoisted at a few stations and consists of a triangle of lights 1.3 metres (4 feet) wide at the base.

If the gale is expected to commence from a northerly point the cone or triangle of lights is hoisted point upwards. If the gale is expected from a southerly point the signal is shown with its point downwards. Gales starting from east or west and likely to change to a northerly direction will be indicated by a north cone whilst those likely to change to a southerly direction will be indicated by a south cone.

The warnings will be taken down only if a period of at least 12 hours is expected to be free of gales.

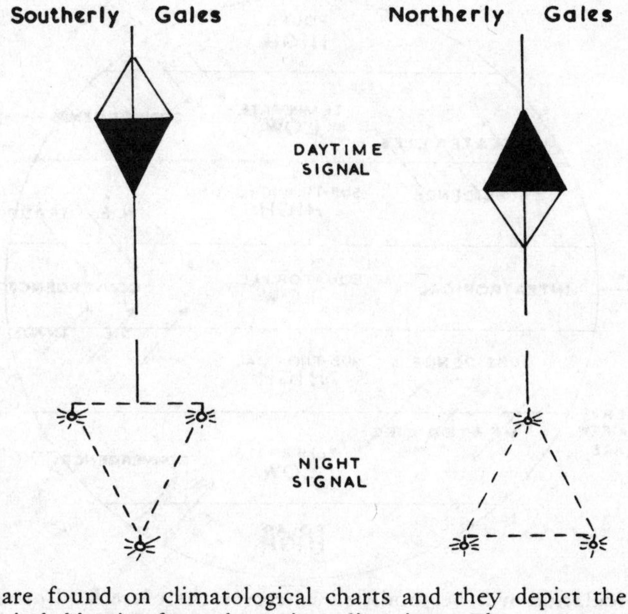

WIND ROSES are found on climatological charts and they depict the frequency and strength of the winds blowing from the various directions. There are many forms of wind rose, each having its own features.

The Baillie rose shown below is comprehensive yet simple.

From the circle to the inner end of the wind arrow represents a scale of 5%. A thin line represents wind force 1 – 3, a double line represents forces 4 – 7, and a broad line forces 8 upwards. The figures in the centre show the number of observations, the percentage frequency of variables and the percentage frequency of calms in that order.

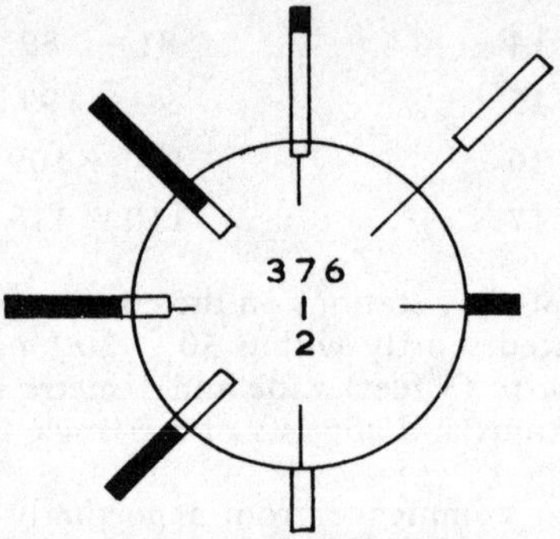

Thus the figure shows North winds 4% light, 10% moderate and 2% gales; NE winds 10% light, 9% moderate; E winds 5% light, 4% gales; S winds 5% light, 5% moderate; SW winds 6% moderate, 7% gales; W winds 2% light, 4% moderate, 9% gales; NW winds 3% moderate, 12% gales. The total number of observations is 376, the frequency of variables is 1% and there is a 2% frequency of calms.

The figure which follows shows the planetary wind circulation for the "ideal" earth which would be one covered all over with water.

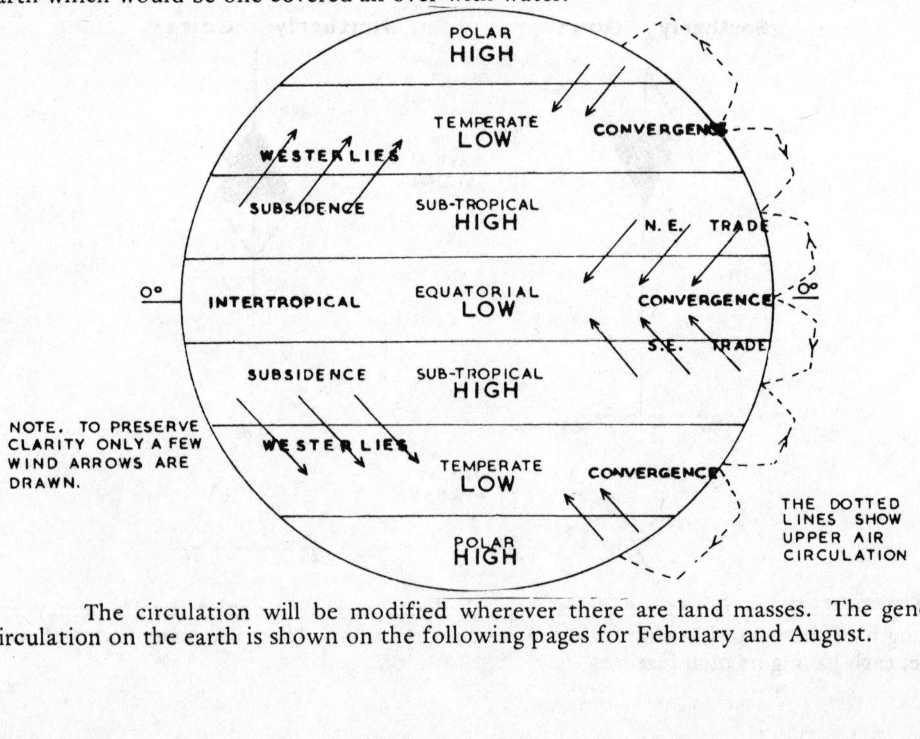

The circulation will be modified wherever there are land masses. The general circulation on the earth is shown on the following pages for February and August.

PREVAILING WINDS & PRESSURE CENTRES

NORTHERN WINTER (DECEMBER to APRIL)

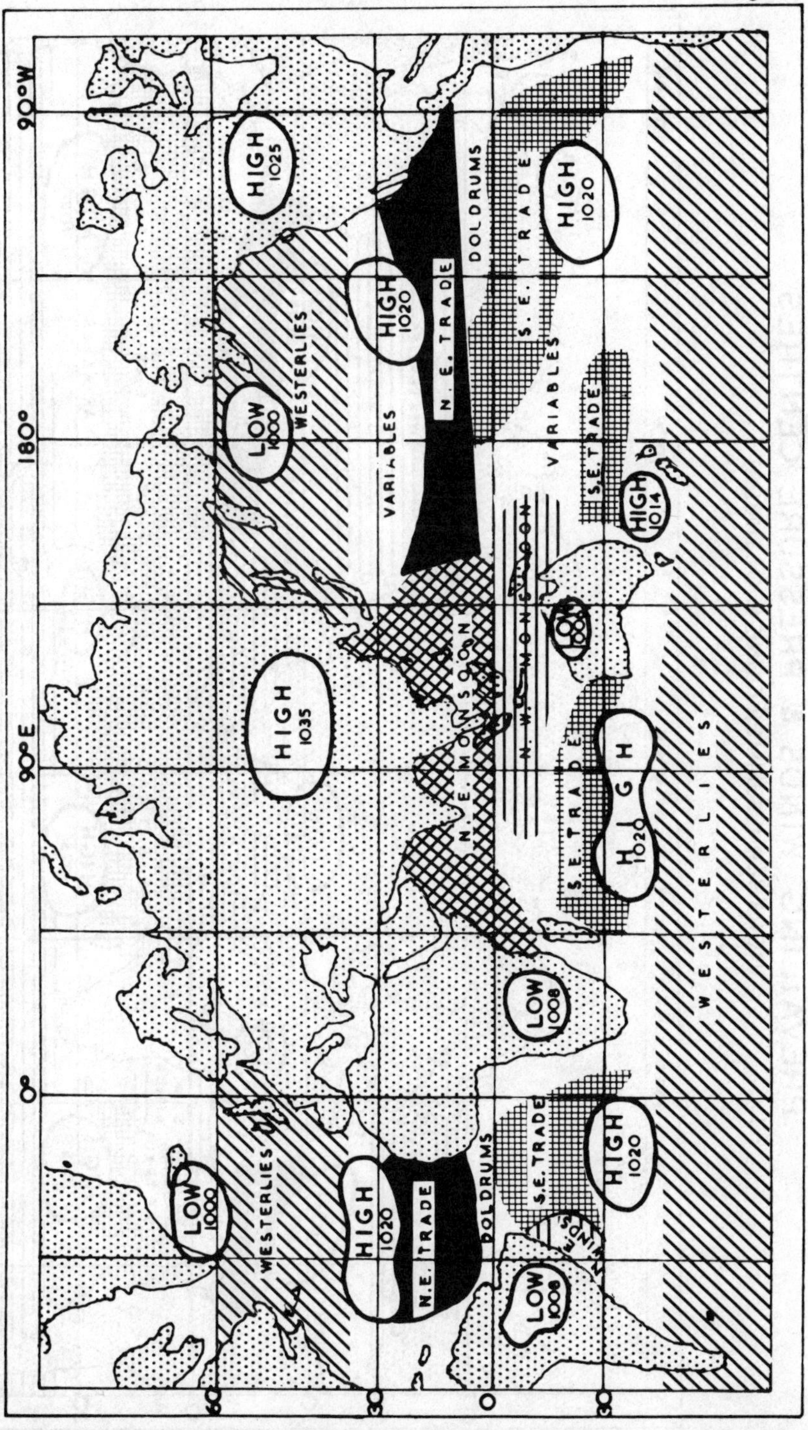

PREVAILING WINDS & PRESSURE CENTRES

NORTHERN SUMMER (JUNE to OCTOBER)

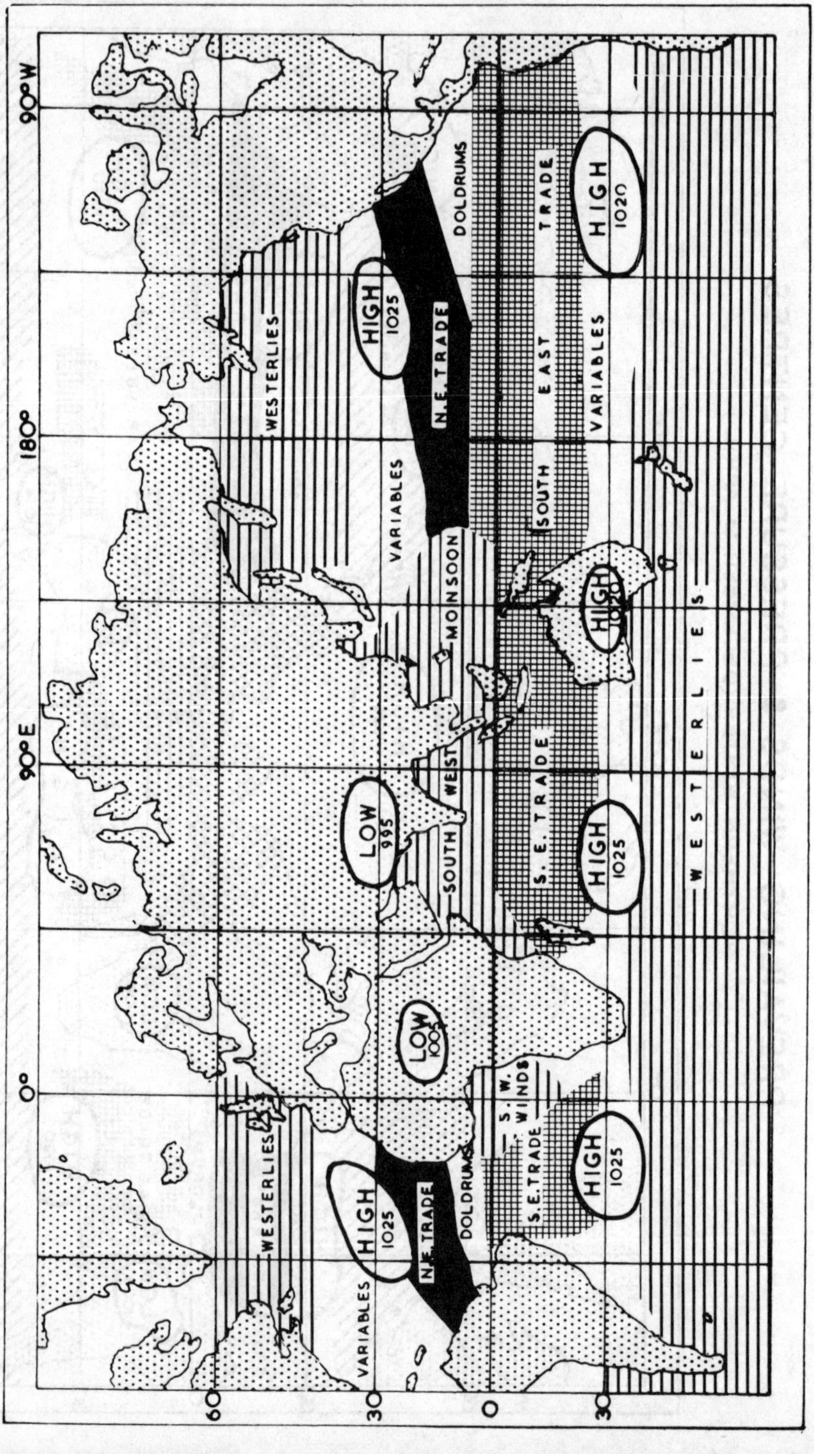

As can be seen the trades move slightly north and south with the sun; their approximate limits are as follows :–

	February	August
Atlantic Doldrums	0° - 2^oN	5^oN - 10^oN
N.E. Trade	2^oN - 25^oN	10^oN - 30^oN
S.E. Trade	0° - 30^oS	5^oN - 25^oS
Pacific Doldrums	4^oN - 8^oN	8^oN - 12^oN
N.E. Trade	8^oN - 25^oN	12^oN - 30^oN
S.E. Trade	4^oN - 30^oS	8^oN - 25^oS
Indian S.E. Trade	15^oS - 30^oS	0° - 25^oS

The ANTI-TRADES are winds which blow above about 2500 metres (8,000 feet) in the opposite direction to the trades on the surface.

LAND AND SEA BREEZES.

As the land is heated during the daytime the air over it will be heated by conduction. This heating causes a decrease in the density of the air and the pressure falls. The sea temperature remains more or less the same and the pressure over it is high compared with that over the land. The pressure gradient is sufficient for air to flow from over the sea to the land; this is the SEA BREEZE.

The sea breeze sets in during the morning, reaches its maximum strength about 1400 hours and then dies away towards sunset.

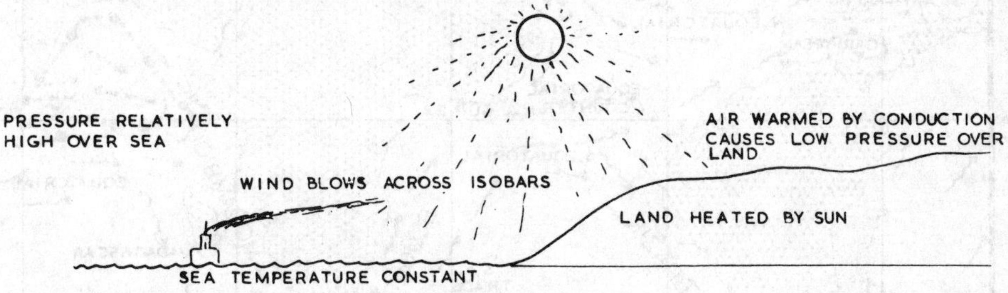

After sunset the land cools rapidly and the air above it also cools and its density increases giving rise to an increase in pressure. The pressure over the sea is now low compared with that over the land. The pressure gradient causes air to flow from the land to the sea; this is the LAND BREEZE.

The land breeze sets in shortly after sunset and continues until dawn.

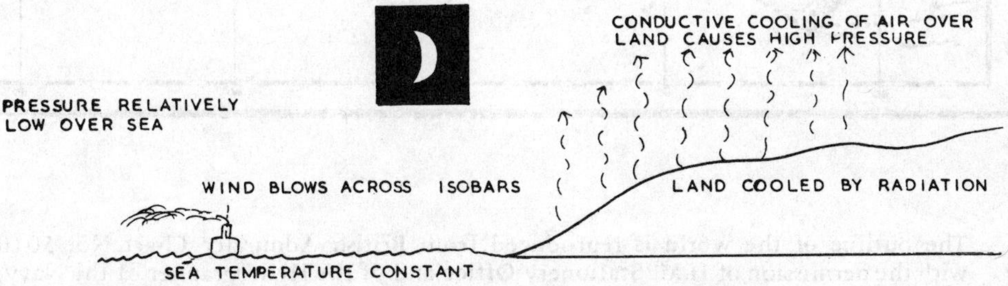

The sea breeze is the more noticeable; it reaches force 3 – 4 and gives a pleasantly cool breeze on the coast during the heat of the day. The land breeze is generally lighter than the sea breeze.

WOR

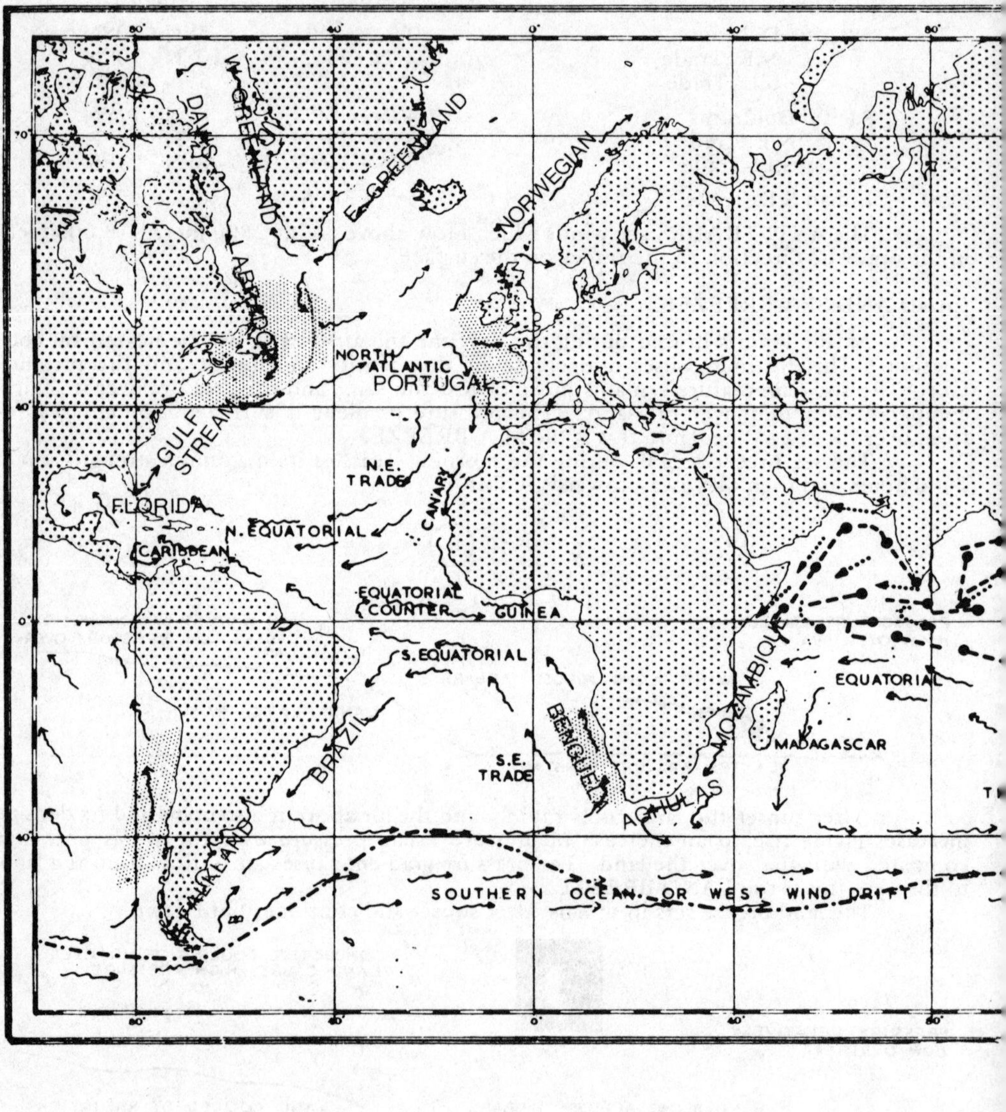

The outline of the world is reproduced from British Admiralty Chart No. 5010 with the permission of H.M. Stationery Office and of the Hydrographer of the Navy.

URRENTS

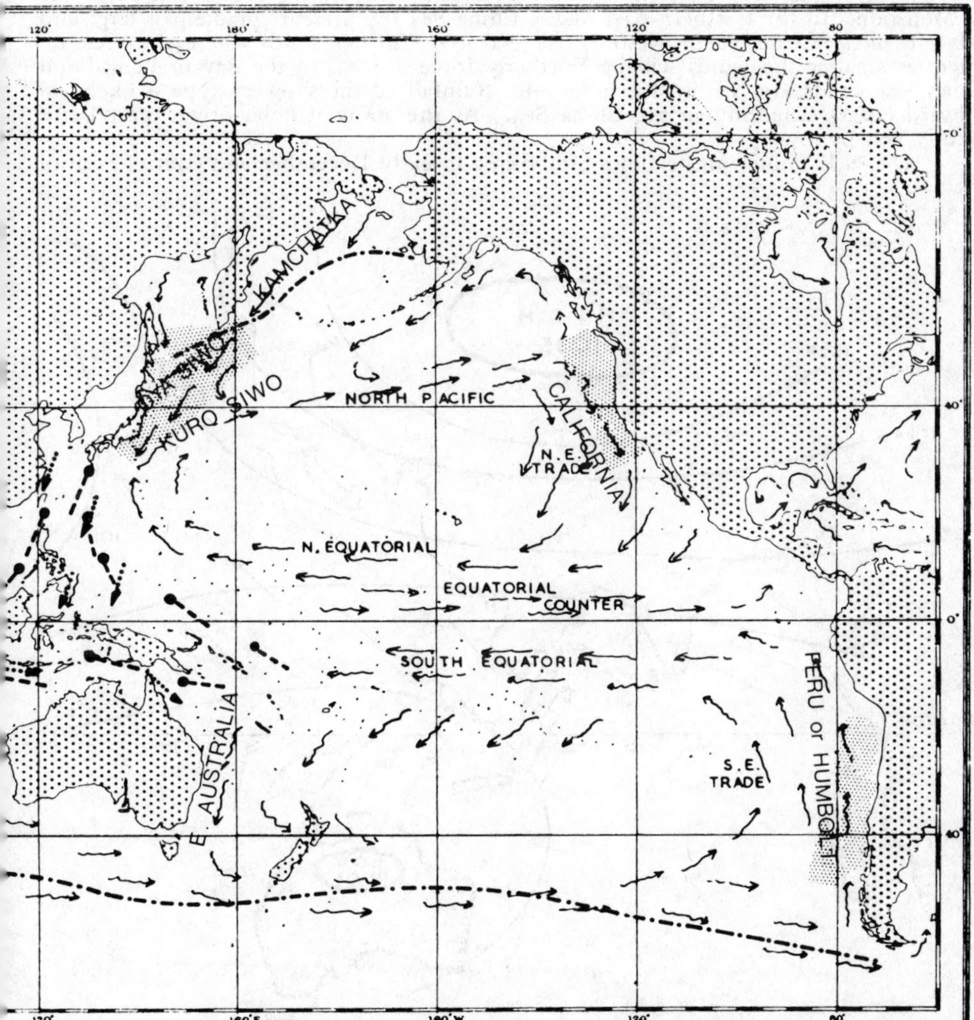
120°
180°
160°
120°
80
70°
40°
0°
40°
120°
160°E
180°W.
120°
60°
KAMCHATKA
OYA SIWO
KURO SIWO
NORTH PACIFIC
CALIFORNIA
N.E. TRADE
N. EQUATORIAL
EQUATORIAL
COUNTER
SOUTH EQUATORIAL
PERU or HUMBOLT
S.E. TRADE
E. AUSTRALIA
GENERAL CURRENTS
S.W. MONSOON CURRENTS
N.E. MONSOON CURRENTS
WARM CURRENTS
COLD CURRENTS
ICE LIMITS
FOG AREAS
S.E. TRADE
KURO SIWO
OYA SIWO

N.E. MONSOON OF THE CHINA SEAS AND INDIAN OCEAN.

During the northern winter the Asian continent is cooled and an intense high pressure area forms over Eastern Siberia. The winds circulating round this form the N.E. Monsoon. In the northern part of the China Sea the pressure gradient is large and winds are likely to be North Westerly force 6 – 7, further south where the pressure gradient is smaller the winds will be Northerly force 5 – 6. In the Bay of Bengal and Arabian Sea winds are N.E.'ly force 3 – 4. Rainfall of the showery type is likely on windward coasts especially in the China Sea. As the air is of Polar origin the relative humidity will be low.

The N.E. monsoon becomes established in early December and continues until April.

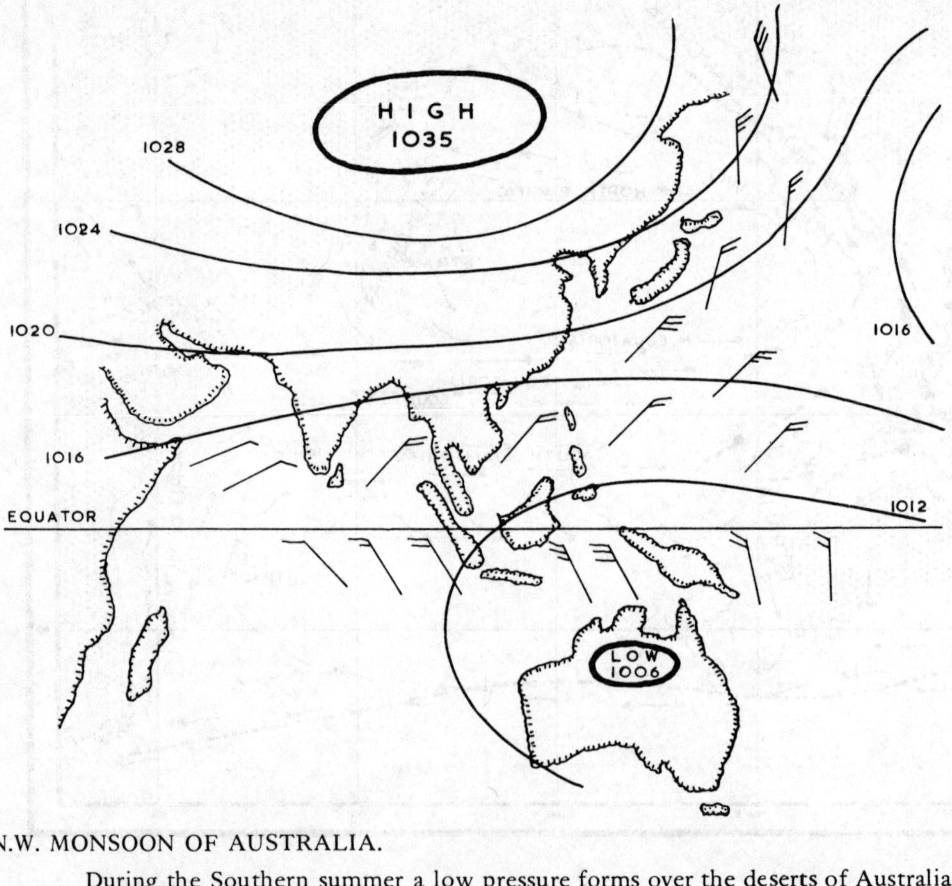

N.W. MONSOON OF AUSTRALIA.

During the Southern summer a low pressure forms over the deserts of Australia. The pressure gradient from the high over Siberia is continued to the aforementioned low. The N.E. monsoon winds blow down to and past the equator. On crossing the equator these winds are deflected to the left giving N.W. winds on the coasts of Australia. Having blown over thousands of miles of ocean the air has a high relative humidity and periods of rain may be expected. The wind force is unlikely to exceed force 6 – 7 and is frequently less. Intensification or otherwise of the Siberian high will affect this monsoon as well as the N.E. monsoon.

The duration of this monsoon is similar to that of the N.E. monsoon.

S.W. MONSOON OF THE INDIAN OCEAN AND CHINA SEA.

During the northern summer the Asian continent is warmed; the high pressure over Siberia declines and is replaced by a low pressure over N.W. India.

There is a pressure gradient between the high pressure over the South Indian Ocean and the lows over Africa and India. This causes the S.E. Trade to blow up to the equator, and on crossing it, the wind near the African coast circulates round the low over Africa and brings rainfall to East Africa. Away from the coast the wind is deflected to the right and blows from a south westerly direction.

In the Arabian sea the wind is S.W. force 7 – 8, in the Bay of Bengal S.W. winds of force 6 – 7 can be expected, whilst in the southern part of the China Sea the wind is S'ly force 4 – 5. The northern part of the China Sea has light S.E.'ly winds.

The air forming the S.W. monsoon has travelled over thousands of miles of ocean and is saturated. This results in heavy rain, especially near the coasts, and poor visibility.

The S.W. monsoon sets in during June and lasts until October, the early months being the worst.

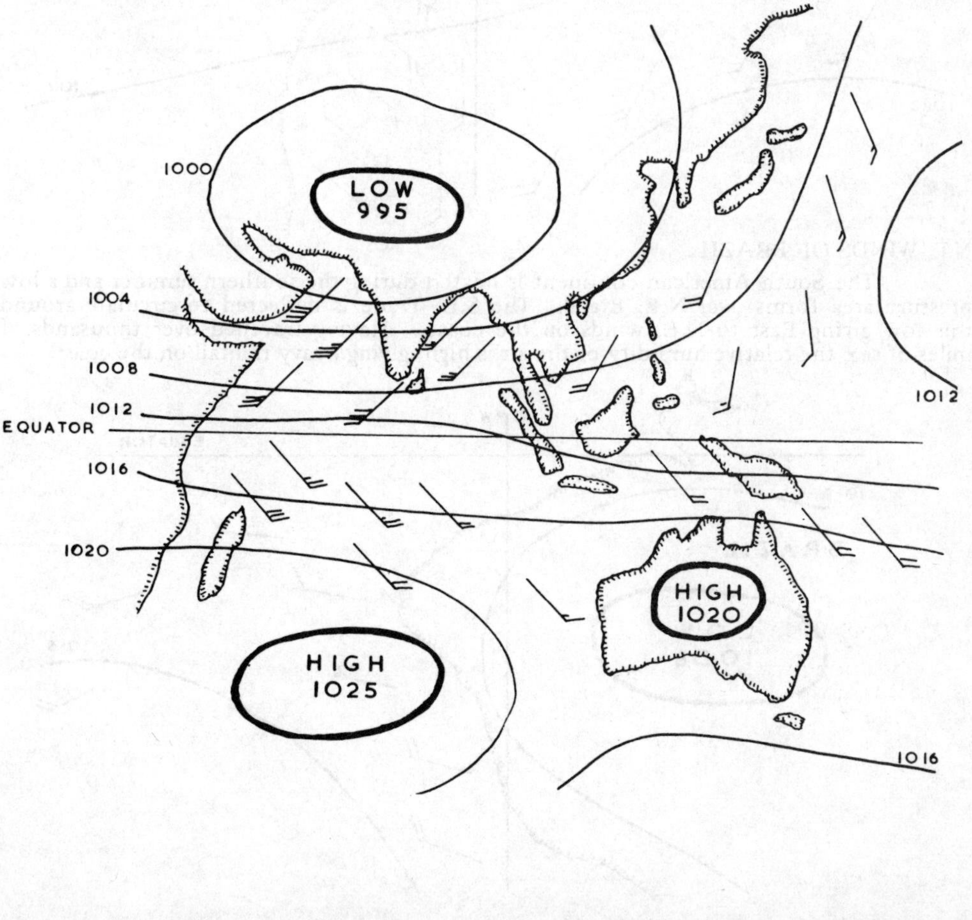

S.W. WINDS OF WEST AFRICA.

During the northern summer the low pressure over Africa moves north of the equator. The pressure gradient between the high over the South Atlantic and the low over Africa causes the S.E. Trade to be deflected and it arrives on the West Coast of Africa as a S.W. wind often called a monsoon, although its intensity is nothing like that of the S.W. monsoon of India.

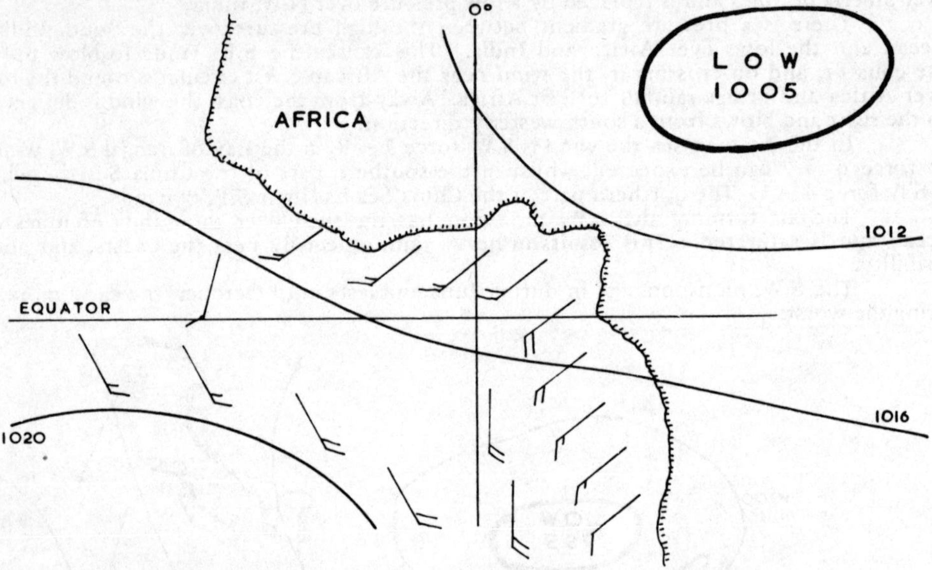

N.E. WINDS OF BRAZIL.

The South American continent is heated during the southern summer and a low pressure area forms over N.E. Brazil. The S.E. Trade is deflected to circulate around this low giving East to N.E. winds on the coasts. Having travelled over thousands of miles of sea, the relative humidity of the air is high, giving heavy rainfall on the coast.

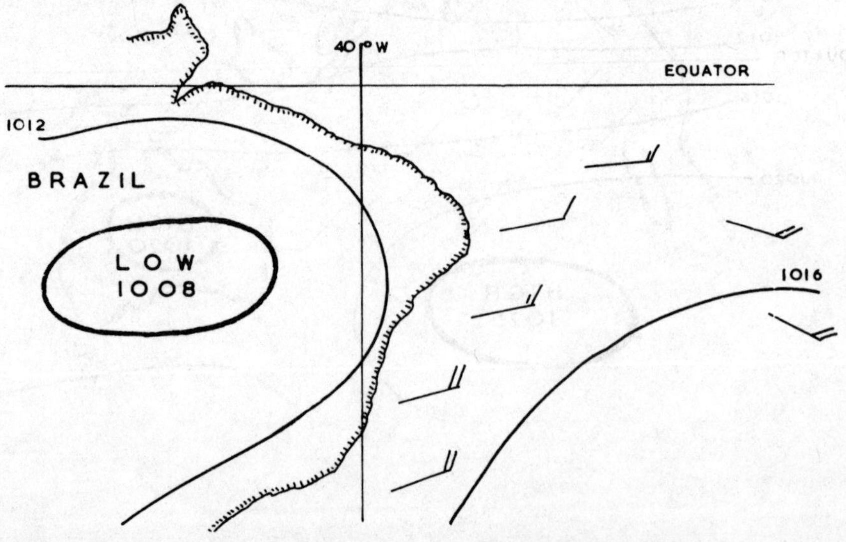

MEDITERRANEAN WINDS

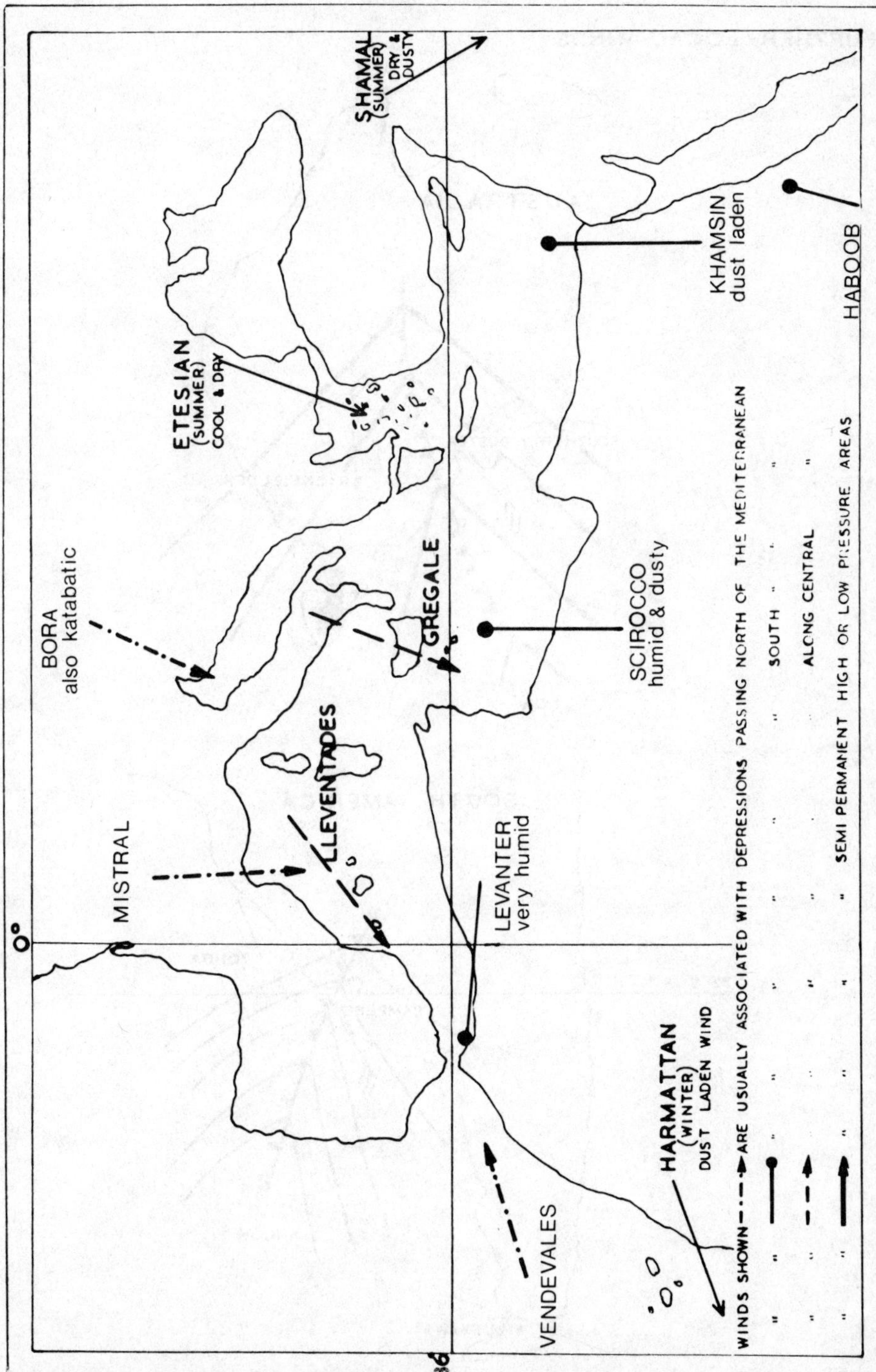

FURTHER LOCAL WINDS

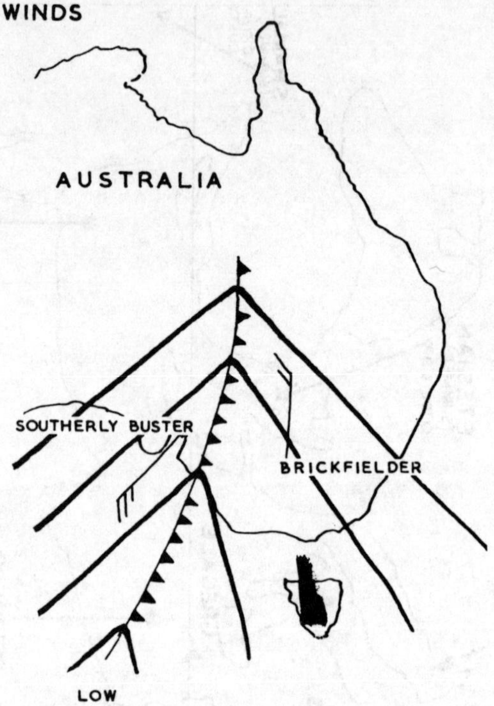

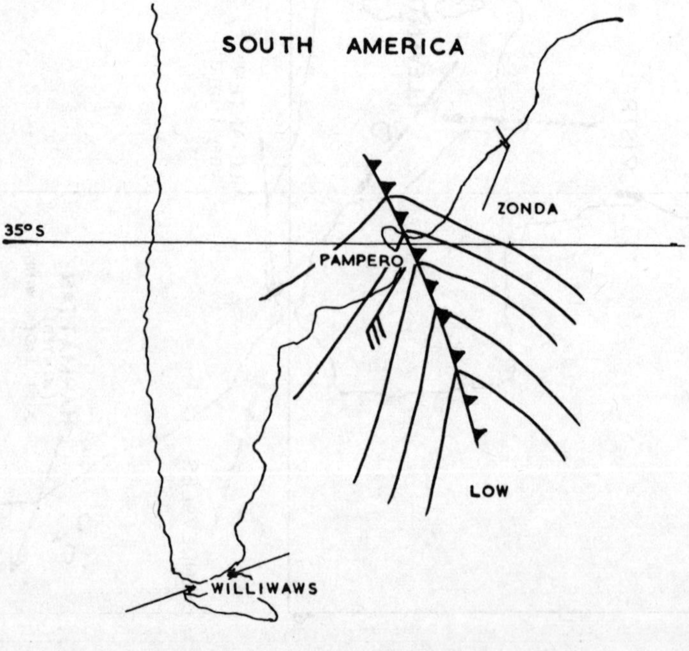

KATABATIC AND ANABATIC EFFECTS.

When the earth's surface cools by night, the air next to the surface is also cooled and thus its density is increased. If this cold air is on high ground there is a tendency for it to sink down to lower ground. If the high ground is a cliff top or a high coastal plateau, the downflowing cold air will move horizontally when it reaches sea level to form a KATABATIC WIND, which may reach force 6–7.

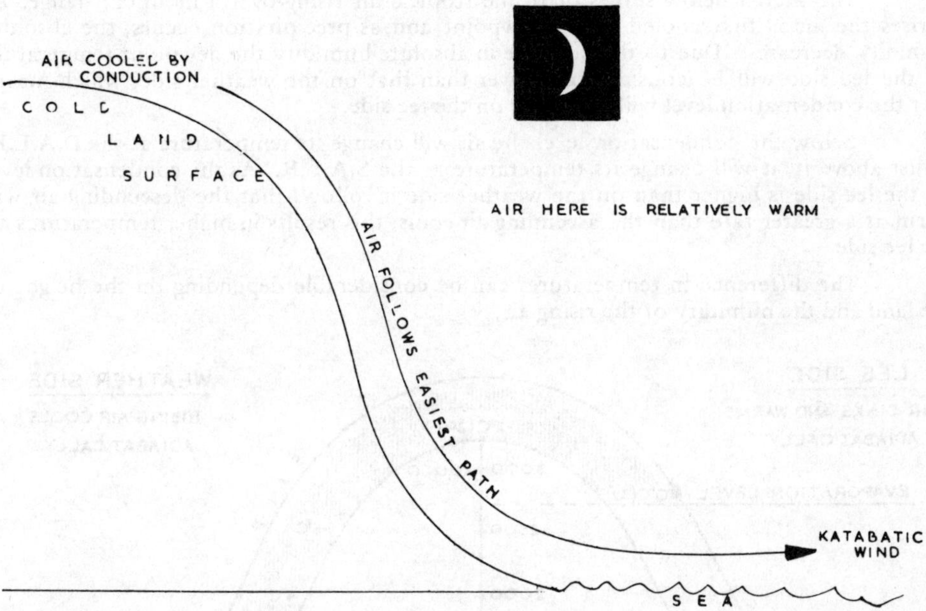

The Katabatic wind is common off the coasts of Greenland where cooling is helped by the snow covered land surface. Also experienced in the Adriatic and many other areas having high land adjacent to the coast.

The ANABATIC WIND is less noticeable as it blows up the sides of valleys with considerably less force than the KATABATIC wind. The air at the bottom of the valley is warmed by conduction from the heated land during the day, and this air, being less dense than the air above it, takes the easiest path to the top of the valley by following the warm sides.

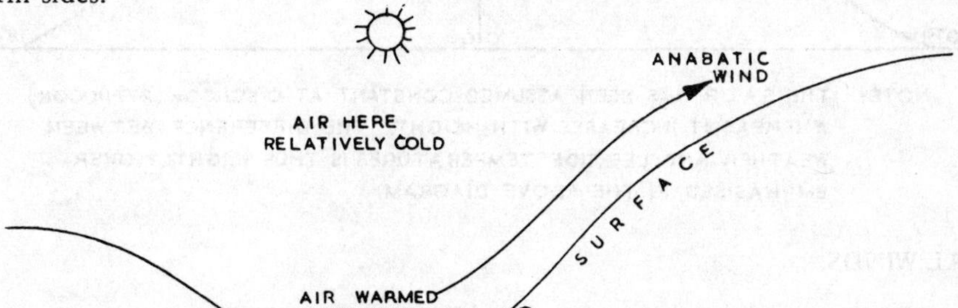

FÖHN WIND EFFECT.

This is felt on the lee side of high ground and results in temperatures on the lee side being higher than those on the weather side if the air passing over the high ground has been cooled to below its dewpoint.

The sketch below shows maritime tropical air rising over a mountain range. As it rises the air is first cooled to its dewpoint and, as precipitation occurs, the absolute humidity decreases. Due to the decrease in absolute humidity the dewpoint temperature on the lee side will be considerably lower than that on the weather side, which means that the condensation level will be higher on the lee side.

Below the condensation level the air will change its temperature at the D.A.L.R. whilst above it, it will change its temperature at the S.A.L.R. As the condensation level on the lee side is higher than on the weather side it follows that the descending air will warm at a greater rate than the ascending air cools; this results in higher temperatures on the lee side.

The difference in temperatures can be considerable depending on the height of the land and the humidity of the rising air.

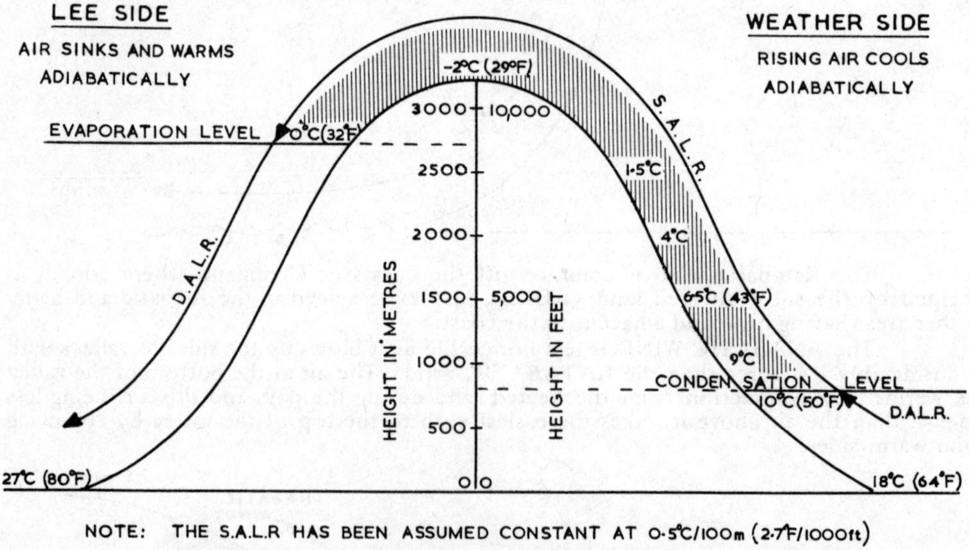

NOTE: THE S.A.L.R HAS BEEN ASSUMED CONSTANT AT 0·5°C/100m (2·7°F/1000ft) WHEREAS IT INCREASES WITH HEIGHT. THE DIFFERENCE BETWEEN WEATHER AND LEE SIDE TEMPERATURES IS THUS SLIGHTLY OVER-EMPHASISED IN THE ABOVE DIAGRAM.

LOCAL WINDS.

In many parts of the world seasonal and occasional winds are given local names; some of the better known of these are illustrated on pages 49 and 50.

V
Isobaric Systems

ANTICYCLONE is the name given to a system of high pressure. The wind circulation around this system will be clockwise in the northern hemisphere and anticlockwise in the southern hemisphere. Anticyclones are associated with small pressure gradients and consequently light winds. As they are areas of subsidence rainfall is unlikely. They may be classed as either cold or warm and each of these may be either permanent (long lasting) or temporary.

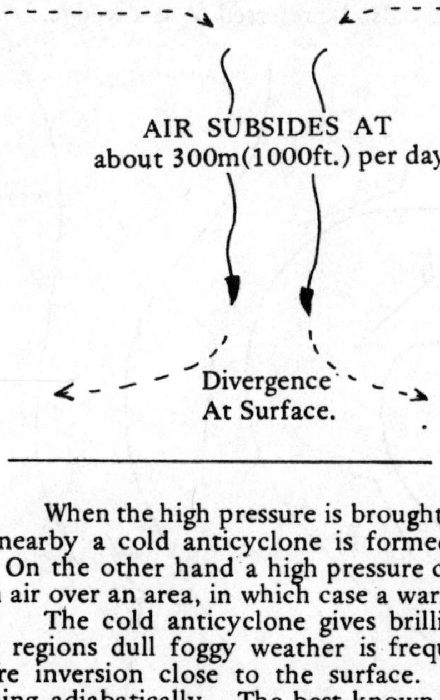

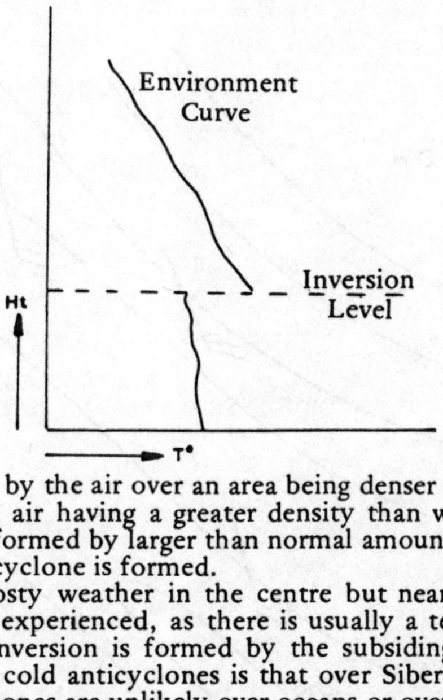

When the high pressure is brought about by the air over an area being denser than that nearby a cold anticyclone is formed, cold air having a greater density than warm air. On the other hand a high pressure can be formed by larger than normal amounts of warm air over an area, in which case a warm anticyclone is formed.

The cold anticyclone gives brilliant frosty weather in the centre but near the outer regions dull foggy weather is frequently experienced, as there is usually a temperature inversion close to the surface. The inversion is formed by the subsiding air warming adiabatically. The best known of the cold anticyclones is that over Siberia in winter which is a permanent one. Cold anticyclones are unlikely over oceans or over the land in summer; if they do form they rapidly become warm ones or disperse.

Temporary cold anticyclones occur between the middle latitude depressions and bring a welcome though brief spell of very bright weather. Occasionally the ridge with which the temporary anticyclone is associated will intensify to build up an anticyclone composed entirely of cold air. If this occurs over the land in winter it may merge with the continental high and be kept as a cold anticyclone by continuous cooling of the land. If it occurs over the land in summer or over the sea at any time it is hardly likely to persist as a cold anticyclone; it may collapse or become a warm anticyclone.

Warm anticyclones occur when the air in the troposphere is warmer than the surrounding air and the high pressure is caused by an excessive depth of this warm air.

Permanent warm anticyclones are those which are found over the oceans and are generally referred to as the sub-tropical highs. They are composed of warm dry air and the visibility is excellent.

Temporary warm anticyclones may occur when there is an extension from the sub-tropical high or when a temporary cold anticyclone warms adiabatically. Over the land the air is very dry and by day fine hot weather will be experienced during the summer. By night there may be sufficient radiation cooling to give fog, this being particularly likely during the autumn.

Over the sea advection fog may occur at any time but this is most probable during spring and early summer.

The pressure tendencies of the anticyclone are frequently helpful in forecasting the movement of depressions, the general rules for these being given on page 62

A RIDGE OF HIGH PRESSURE is an extension of an anticyclone and may be in any direction from its parent. During the winter time ridges from the sub-tropical high and from a high over Greenland frequently join to give a continuous period of Northerly winds over the British Isles(see page 67). The ridge may also be referred to as a wedge.

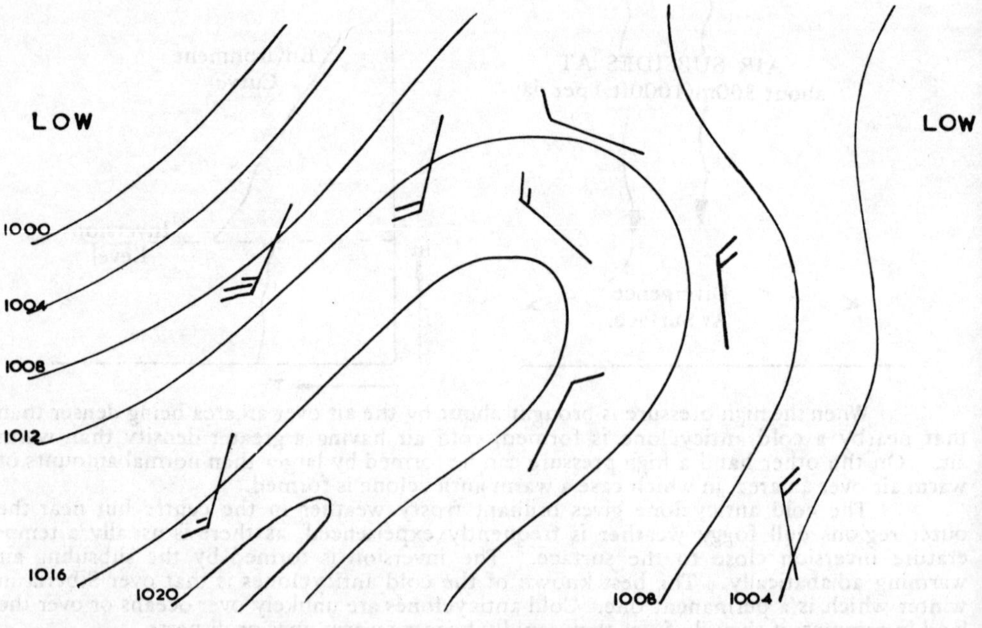

AIR MASSES are large areas of air which have the same characteristic throughout. For such conditions to occur the air must remain over an area for several days. The ideal conditions will be found where there is a permanent anticyclone and these are the principal SOURCE REGIONS for the air masses.

Air masses may be either warm or cold depending on whether their source is tropical or polar. They are further described as being either maritime or continental.

The following table and the diagram describe and show the approximate tracks of air masses which affect the British Isles.

AIR MASS AND ABBREVIATION	SOURCE	TEMPERATURE	RELATIVE HUMIDITY	SKY	AIR CONDITION
Maritime tropical mT	Azores	High	High	Covered with stratus	Stable
Continental tropical cT	Sahara	Very high	Low	Clear	Stable
Maritime polar mP	Greenland	Low-high depending on time over sea	Medium	Cumulus cloud	Unstable
Continental polar cP	Siberia	Very low	Low	Possible stratus through turbulence	Stable
Arctic A	Arctic Ocean	Very low	low - medium	Cumulus	Unstable

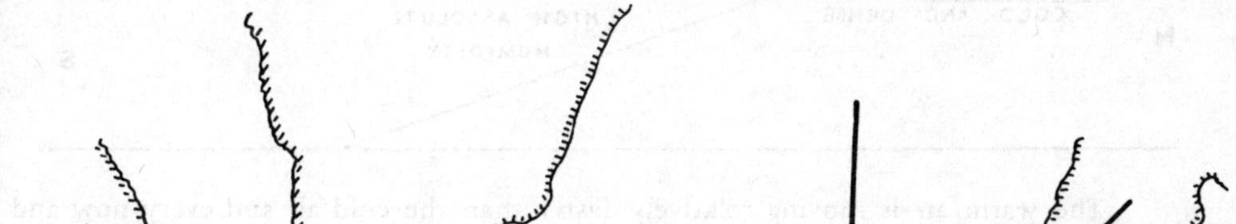

It can be seen that each air mass has a different characteristic to the other air masses and consequently they will not mix with one another. The boundary of each air mass is generally well defined and on the ground it is known as a FRONT. The boundary above the ground is known as a FRONTAL SURFACE. Most fronts are in motion, but if a front is stationary or behaving as if it were stationary, it is referred to as being QUASI-STATIONARY.

When the approaching air is colder than the present air the boundary will be a COLD FRONT whereas if the approaching air is warmer than that at present the boundary will be a warm front.

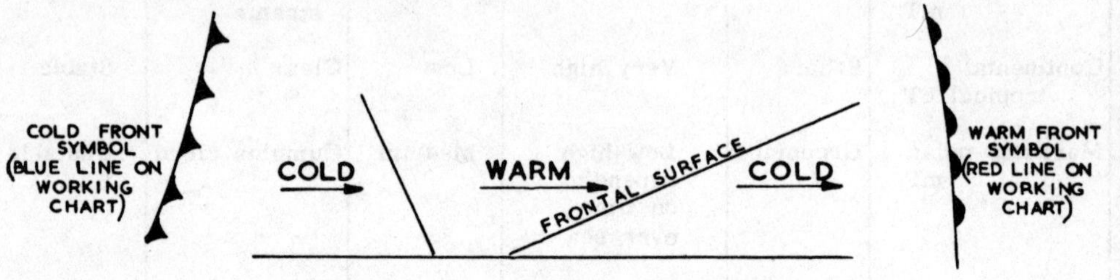

The general line of demarcation between polar and tropical air masses is known as the POLAR FRONT. The density of the polar air is greater than that of the tropical and the frontal surface cannot be vertical, but is inclined polarwards so that the tropical air overrides the polar air. The slope of the frontal surface is about 1 in 150.

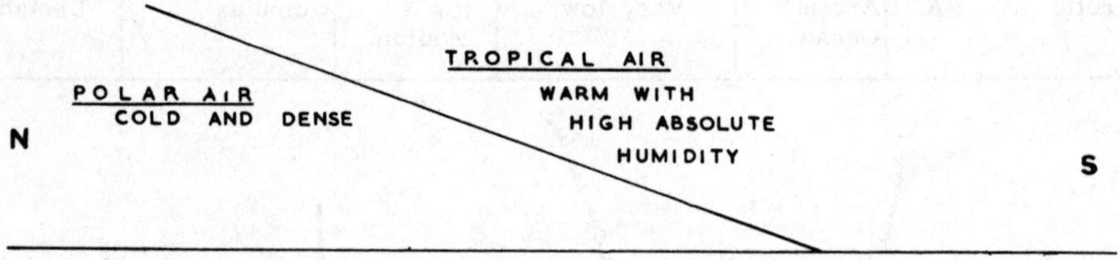

The warm air is moving relatively faster than the cold air and every now and again it forces its way into the cold air making a bulge in the polar front. This is the commencement of a FRONTAL DEPRESSION, for as the warm less dense air pushes into the cold denser air the pressure falls. The sketches show the successive stages in the formation and degeneration of the depression.

It will be seen that when the depression is fully developed the air behind the cold front is moving faster than the air in the warm sector. This causes the warm sector to be undercut and lifted off the ground; the front is then said to be OCCLUDED.

The air ahead of and behind the occlusion was originally of the same temperature but circumstances (the warm sector intervening and the passage of time) cause a change in their temperatures. If the air behind the occluded front is colder than that in front the occlusion is said to be a cold one, but if the air behind the occlusion is warmer than that ahead of it, the occlusion is a warm one.

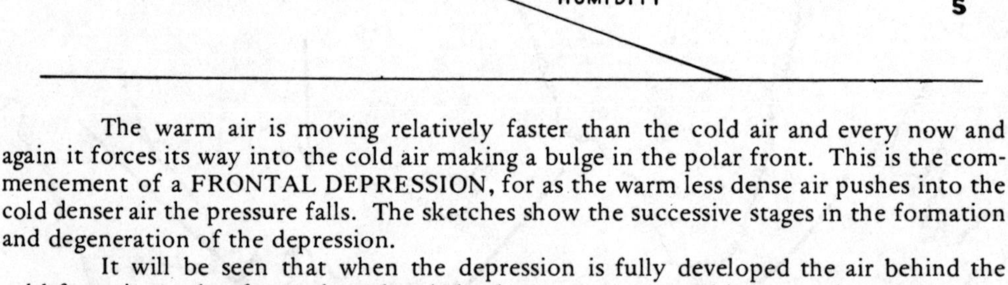

SKETCHES ILLUSTRATING THE LIFE OF A FRONTAL DEPRESSION

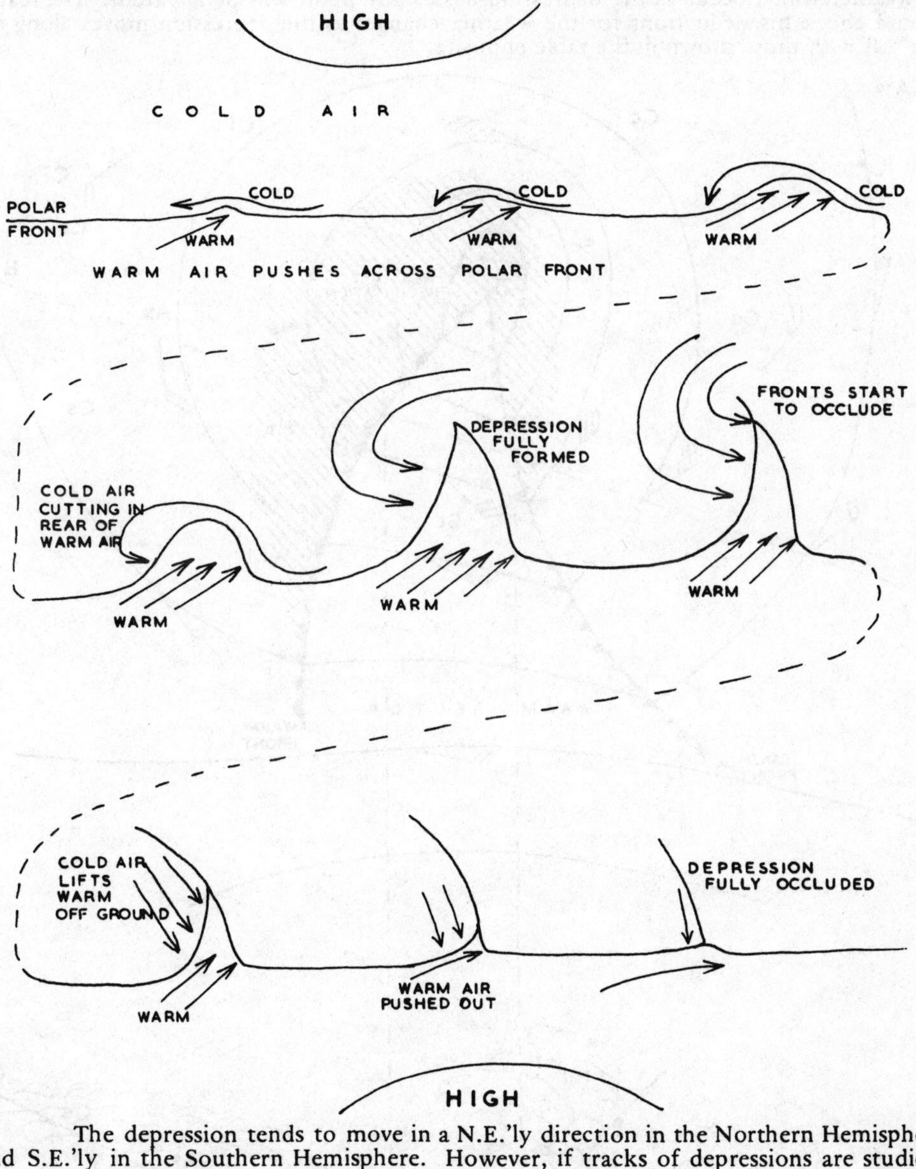

The depression tends to move in a N.E.'ly direction in the Northern Hemisphere and S.E.'ly in the Southern Hemisphere. However, if tracks of depressions are studied, they will be seen to be in every direction, although the above directions predominate. Temperate latitude depressions moving westwards are rare and even if they do move in this direction, they will not do so for any length of time.

If the way in which the depression is formed is considered carefully the changes of weather which occur as the depression passes any point will be apparent. The reader should check his deductions for the weather changes as the depression moves along the line AB with those shown in the table opposite.

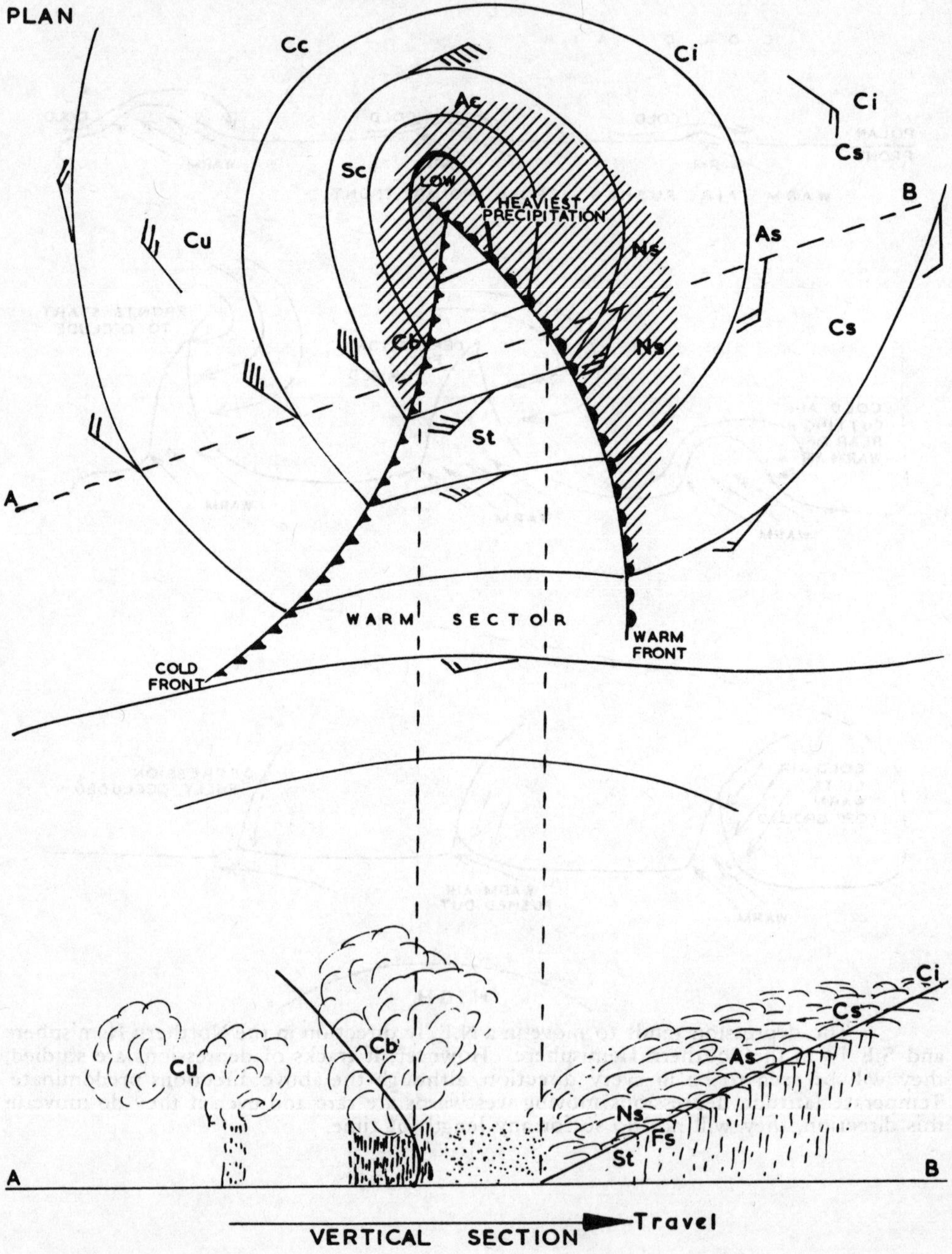

FRONTAL DEPRESSION
(WEATHER CHANGES)

	WARM FRONT		WARM SECTOR	COLD FRONT	
	APPROACHING	PASSING		PASSING	AT THE REAR
PRESSURE	Steady fall	Stops falling	Little change	Sudden rise	Rising steadily
WIND	Veers from S-SW and increases	Veers with possible increase in velocity	Steady, possibly backing as cold front approaches	Sudden veer from SW-W or NW with squalls	Velocity tends to decrease, steady in direction
TEMPERATURE	Rising slowly	Slight rise	Little change with relatively high temperatures	Sudden drop	Little change but tendency to fall.
SKY	Becoming overcast Ci, Cs, As, Ns.	Ns and Fs	Overcast with St turning to Sc	Cb	Cb, Ac and Cu. With blue sky.
PRECIPITATION	Continuous from drizzle to heavy rain or snow	Rain stops but may be slight drizzle	Intermittent slight rain or drizzle possibly fog.	Heavy rain, thunder and possibly hail	There may be a narrow belt of continuous heavy rain, turning to heavy showers later.
VISIBILITY	Deteriorating	Poor	Poor	Great improvement	Excellent except in showers.
HUMIDITY	Increasing	Rapid rise	Very high	Rapid fall	Fairly low but variable in rain
UPPER AIR	Stable	Very stable	Stable	Very unstable	Unstable

A southern hemisphere frontal depression is formed in exactly the same way as one in the northern hemisphere – namely by a warm maritime air mass encroaching on a cold polar air mass. The reader should attempt the drawing of a formation of such a depression and then compare his fully developed depression with the plan view shown below.

The section along AB is exactly the same as that shown on page 58 and the weather changes on page 59 are also applicable except that for the wind. The wind will back from N.W.'ly to S.W.'ly as the depression approaches and passes. It should be noted that the feathers on the wind arrows in the southern hemisphere should always be placed on **the left hand side** of the wind arrow when facing the wind.

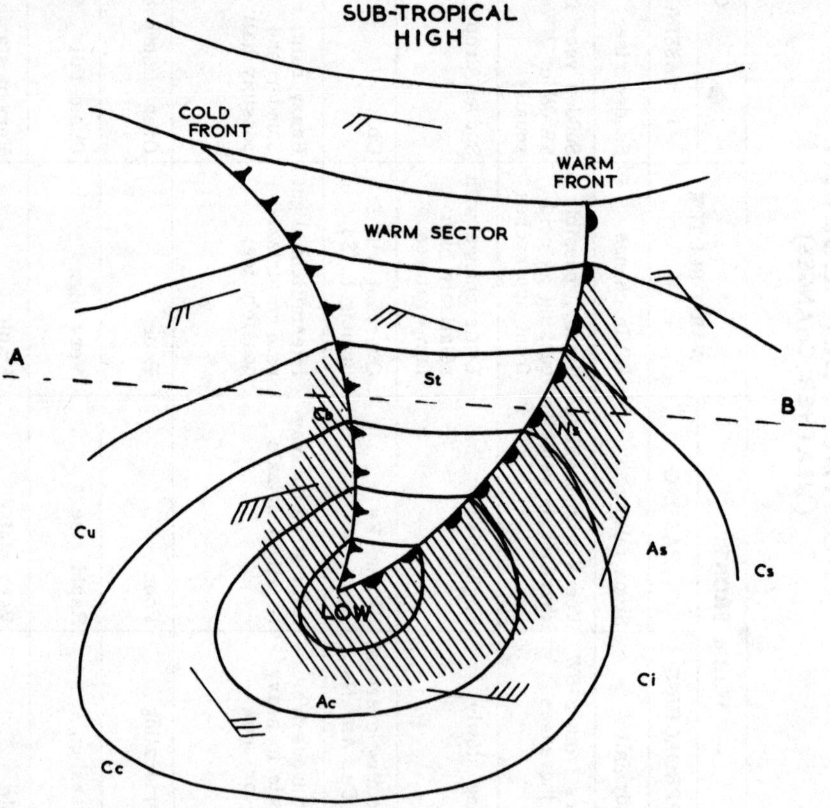

PLAN OF SOUTHERN HEMISPHERE
FRONTAL DEPRESSION

FRONTOLYSIS is the term applied to "spreading" of the isotherms at a front which means that the front is less active and the depression will be filling. This occurs when the isobars are cyclonic.

FRONTOGENESIS is the term applied to the "closing" of the isotherms at a front. This means that the activity at the front is increasing with a deepening of the depression or more likely the formation of a secondary depression. It occurs when the isobars are anticyclonic.

The ideal conditions for the formation of a SECONDARY DEPRESSION are usually found on the cold front between the point where the warm and cold fronts are occluded and the point where the front runs parallel to the isobars. The secondary, which may only show as a bulge in the isobars, moves around the primary (the original depression) in an anticlockwise direction and tends to deepen as the primary fills. The phenomena associated with a secondary are often more violent than those around the primary.

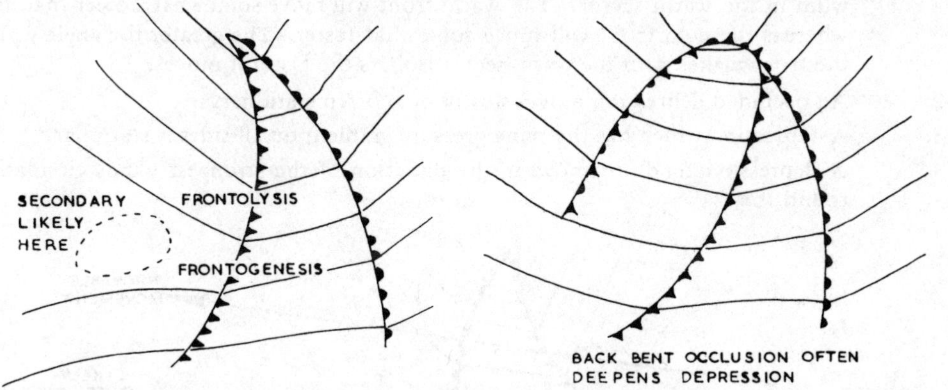

Secondary depressions may form where there is a kink or wave on the cold front; they may then be referred to as WAVE DEPRESSIONS.

As the cold front of a frontal depression 'trails' to join up with the polar front it is quite likely that when the next depression is already forming on the polar front, the trailing cold front will join the warm front of the new depression. This may be repeated with several depressions, and is known as a FAMILY OF DEPRESSIONS which may be six or seven strong.

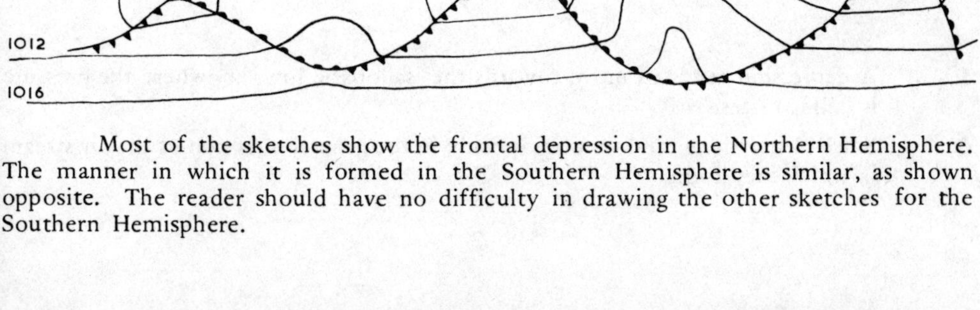

Most of the sketches show the frontal depression in the Northern Hemisphere. The manner in which it is formed in the Southern Hemisphere is similar, as shown opposite. The reader should have no difficulty in drawing the other sketches for the Southern Hemisphere.

Depressions may also be of non-frontal origin, these being formed when there is either a large amount of surface heating or in the lee of mountain ranges or where there is great vertical instability.

The latter type occur particularly in polar air streams; the depression thus formed may remain small or it may grow to quite a large size.

A knowledge of the movement of depressions is essential if any attempt is to be made to forecast the weather. Although each depression must be treated on the merits of the general synoptic situation the following general rules can usually be applied.

1. A frontal depression tends to move in a direction parallel to the isobars in the warm sector. It moves at a speed approximately equal to that of the geostrophic wind in the warm sector. The warm front will move somewhat slower than this whereas the cold front will move somewhat faster. The greater the angle which the front makes with the warm sector isobars the faster it moves.
2. An occluded depression moves slowly or is often stationary.
3. A depression which has the same pressure gradient on all sides is stationary.
4. A depression tends to move in the direction of the strongest winds circulating round it.

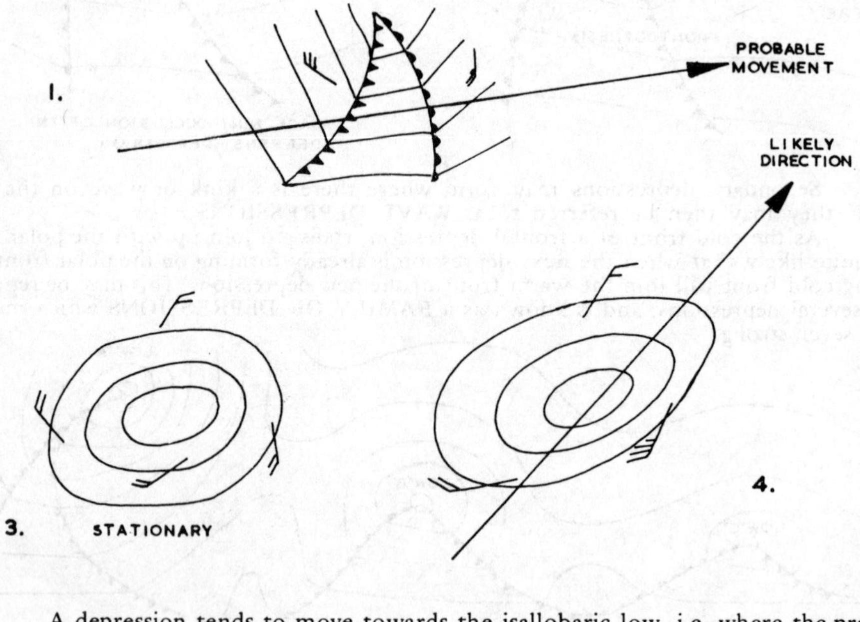

5. A depression tends to move towards the isallobaric low, i.e. where the pressure is falling fastest.
6. Small depressions tend to move quickly in the general direction of the air stream in which they form.

7. Secondary depressions move cyclonically around their primary and if two depressions are more or less the same size forming a dumb-bell shape these tend to revolve cyclonically about a common centre somewhere between the two.

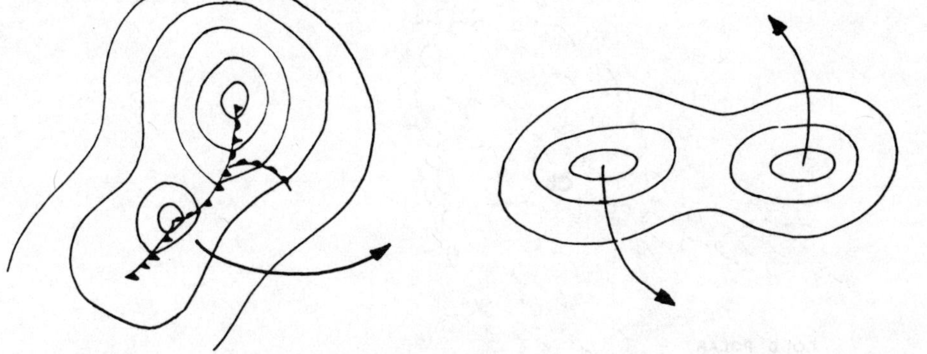

8. Elongated or elliptical shaped depressions tend to move in a direction between the direction of the major axis and the direction of the isallobaric low.

9. Depressions having wind less than normal tend to deepen and those with winds greater than normal tend to fill.

A TROUGH OF LOW PRESSURE, sometimes called a V-shaped depression, is an extension of a depression into a high pressure area. It nearly always points towards the equator. The trough may be frontal, in which case there is a marked change in the direction of the isobars on the trough line, or non-frontal where the isobars are well rounded.

Frontal troughs may be associated with warm, cold, secondary cold or occluded fronts and the weather changes will be similar to those associated with these fronts.

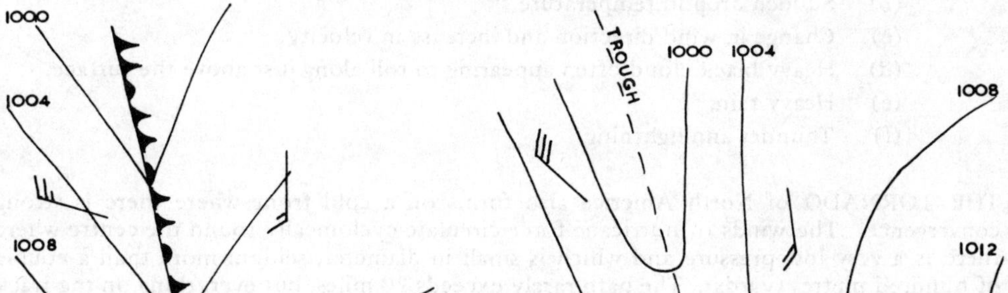

FRONTAL NON-FRONTAL

A LINE SQUALL occurs when the cold front is well marked. In other words, there is a large temperature difference between the warm and cold air. The line squall is common off the East Coast of South America during the spring where it is known as a PAMPERO. In South Eastern Australia a similar squall is known as a SOUTHERLY BUSTER.

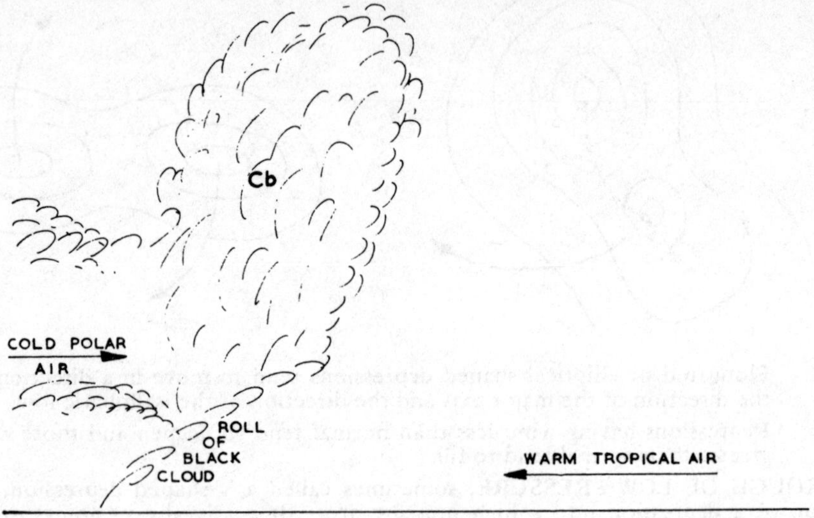

The following changes in the weather may be expected at the passage of a line squall.

(a) Sudden rise in barometric pressure.
(b) Sudden drop in temperature
(c) Change in wind direction and increase in velocity.
(d) Heavy black cloud often appearing to roll along just above the surface.
(e) Heavy rain.
(f) Thunder and lightning.

THE TORNADO of North America also forms on a cold front where there is strong convergence. The winds of hurricane force circulate cyclonically round the centre where there is a very low pressure and which is small in diameter, seldom more than a couple of hundred metres (yards). The path rarely exceeds 20 miles, but everything on the track is likely to have suffered great damage:— roofs ripped off, trees uprooted and windows blown out. Similar tornadoes occur infrequently in England.

A similar but smaller phenomenon over the sea forms a WATERSPOUT. The low pressure and the convergence cause the sea to be sucked up towards the lower portion of the cumulonimbus cloud from which a funnel shaped cloud goes down towards the sea.

The tornadoes of West Africa are very different from those of North America, in that they are violent squalls of wind, without any cyclonic circulation, blowing outward from an approaching thunderstorm. They occur mainly during the rainy season.

When two anticyclonic systems and two cyclonic systems are diametrically opposed there is an area in the centre of the four systems which cannot be considered as a high pressure or a low pressure area. This area is known as a COL and the pressure is lower than that round the high pressures and higher than that round the low pressures.

In a Col the pressure gradients are small giving light variable winds. In general the relative humidity is high and there may be fog, or there may be thunderstorms.

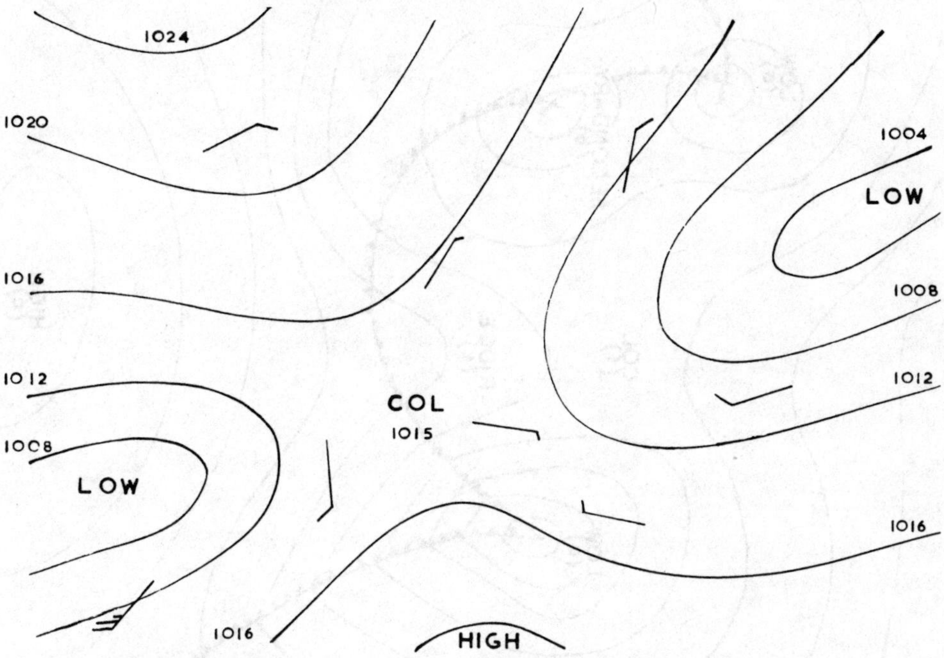

When the distance from a centre of high or low pressure is large the isobars may run in parallel straight lines over a large area, a situation which is known as STRAIGHT ISOBARS.

The weather associated with straight isobars is usually of the changeable type, although in certain situations a spell of fine weather can be expected.

The sketch on page 66 shows pictorially the following principal isobaric forms.

(a) Cyclonic
(b) Secondary
(c) Trough
(d) Anticyclonic
(e) Ridge
(f) Col
(g) Straight isobars.

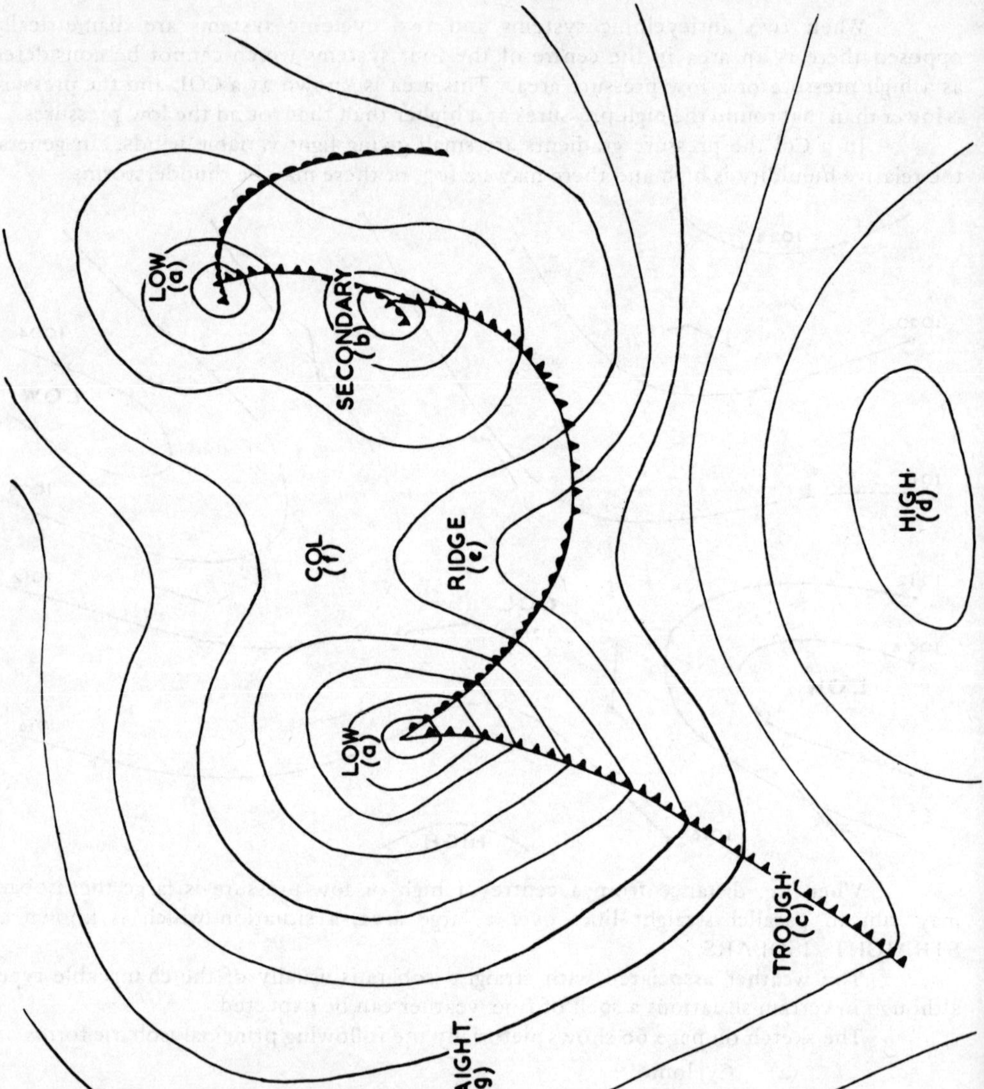
LOW (a)
SECONDARY (b)
COL (f)
RIDGE (e)
HIGH (d)
LOW (a)
TROUGH (c)
STRAIGHT (g)
HIGH (d)

The disposition of anticyclones around the British Isles gives rise to Northerly, Easterly, Southerly, and Westerly types of weather. It is emphasised that a northerly wind does not of itself indicate a Northerly type of weather; there must also be the right disposition of high and low pressure areas.

NORTHERLY TYPE occurs when a high pressure over the Greenland—Iceland area has extensions towards the Azores high, with relatively low pressure over Scandinavia. Most likely in Spring and early Summer. The weather is cold and bright with showers of rain, sleet or snow on high ground and exposed eastern coasts.

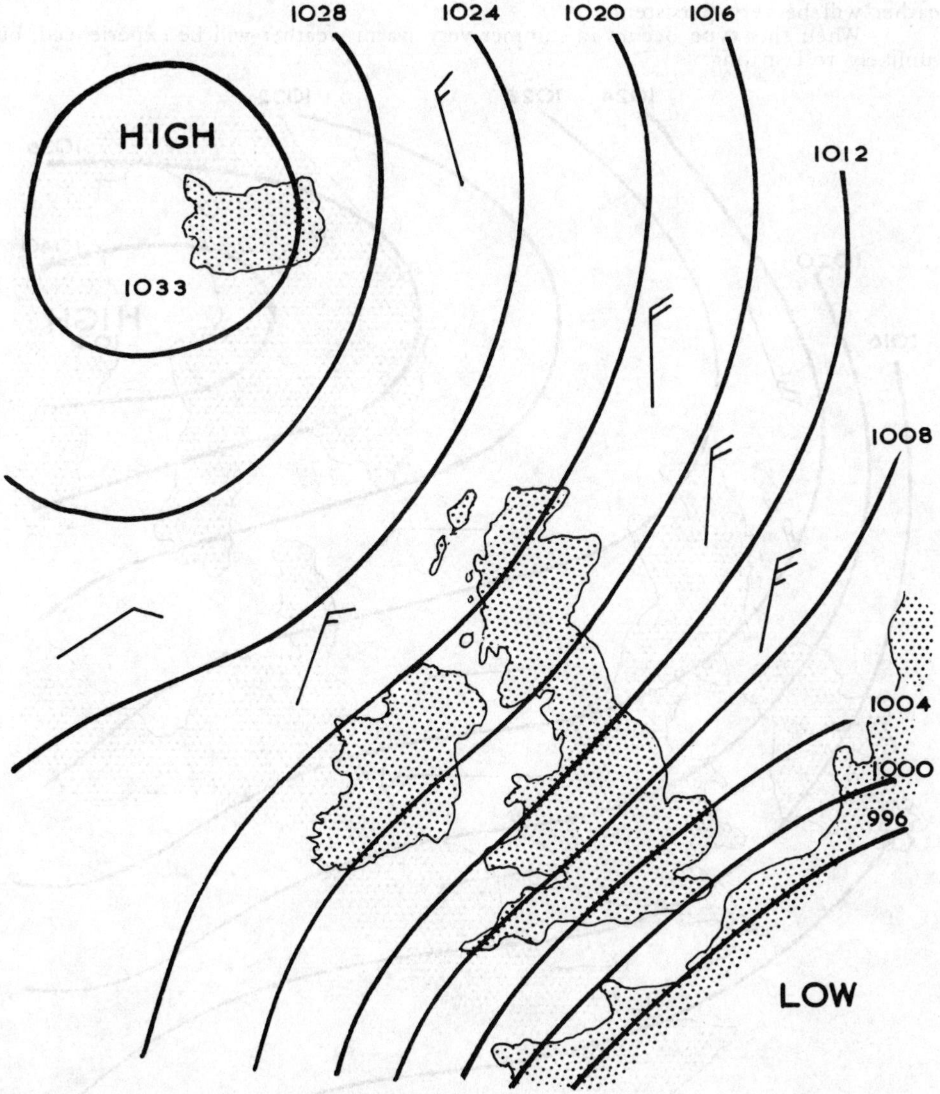

EASTERLY TYPE occurs when there is a high pressure over Scandinavia which extends towards Iceland. Most likely in winter and spring when very cold easterly winds flow over the continent and to the British Isles. There is heavy frost by day and night. By night skies are usually clear whilst by day a layer of turbulent cloud (low stratus) frequently covers the sky.

Due to the Atlantic depressions being prevented from following their usual track up to the Norwegian Sea, they frequently move south to pass over France where the pressure is somewhat lower. As these depressions pass to the south of the British Isles, easterly gales may be experienced together with prolonged snowfall over Southern England. If the Scandinavian high is joined to the Siberian high the Easterly type weather will be very persistent.

When this type occurs in summer very warm weather will be experienced, but is unlikely to last long.

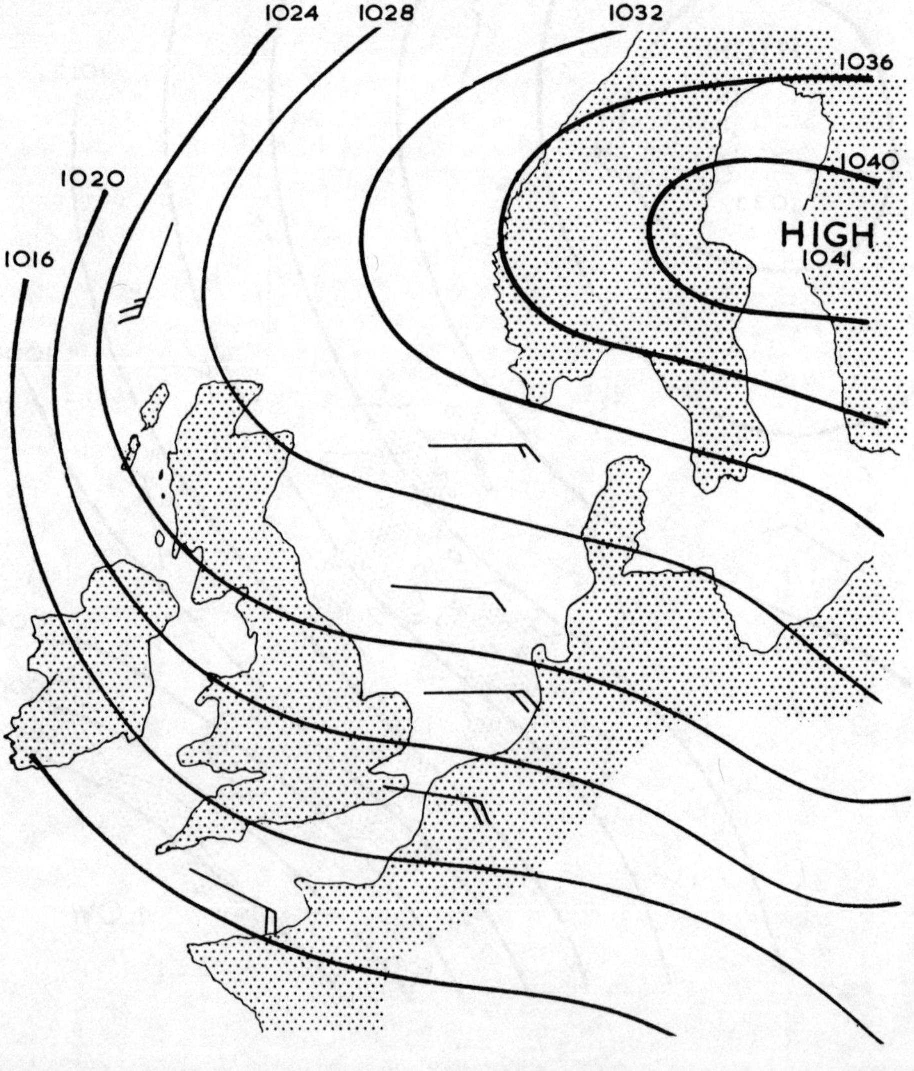

SOUTHERLY TYPE occurs mainly during the autumn and early winter when there is a high pressure over the continent. The British Isles experiences a warm, humid southerly to south westerly airstream, which is associated with the warm sectors of depressions on N.N.E.'ly tracks.

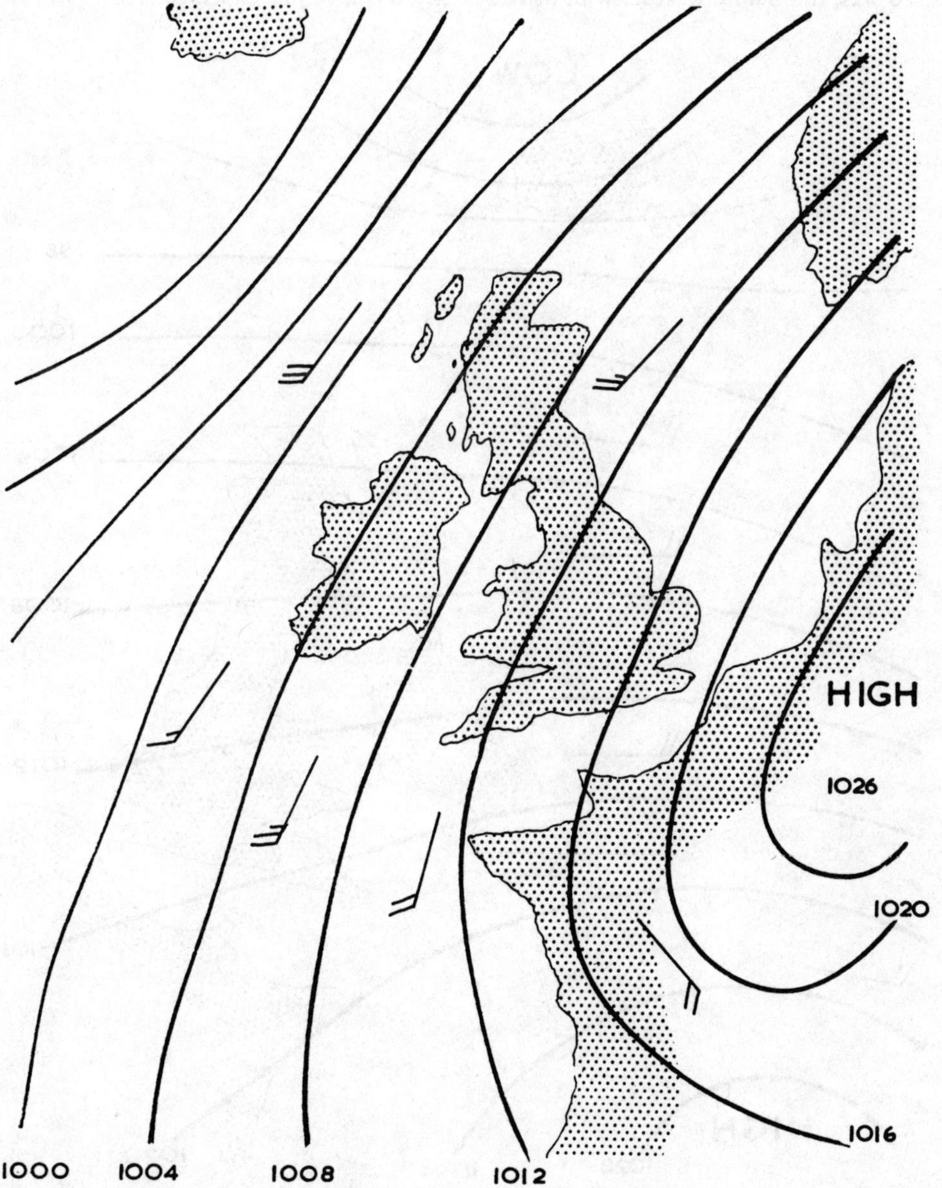

WESTERLY TYPE may occur at any time. In winter changeable weather may be expected as the high pressure is situated well to the south over the Azores, and the weather is influenced by passing depressions.

In summer when the subtropical high pressure area is often situated well north of the Azores, fine sunny weather with light westerly winds can be expected.

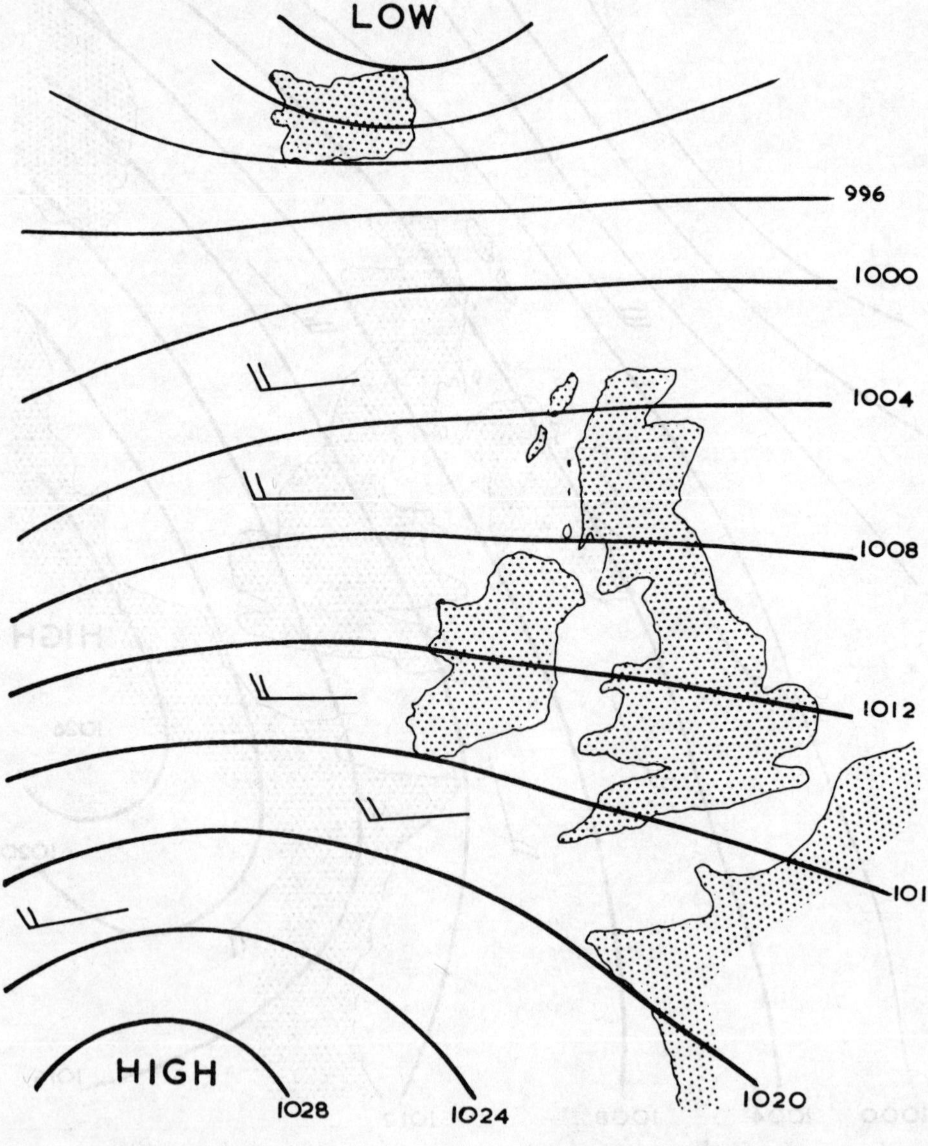

VI
Tropical Storms

It will be remembered that the air circulations of the Northern and Southern hemispheres are in opposite directions, so that disturbances which form in one hemisphere cannot cross into the other hemisphere.

The area near the equator is one of convergence as the Northern Hemisphere N.E. Trade wind blows towards it, as does the S.E. Trade of the Southern Hemisphere. The general name for this area is the INTERTROPICAL CONVERGENCE ZONE (ITCZ), also known as the DOLDRUMS.

When the ITCZ is well to the north or south of the equator the change in direction of the trades after crossing the equator will cause very strong convergence currents and it is possible that a cyclonic disturbance will form in this area. The possibility is increased when the ITCZ is in the vicinity of islands when local surface heating of air of high humidity gives rise to very unstable conditions.

Low pressure areas frequently occur in the ITCZ but cyclonic circulations can only result if the geostrophic force is sufficiently large (there is no geostrophic force on the equator) and this is unlikely in latitudes less than 5°.

The cyclonic disturbance, once formed, is known as a TROPICAL REVOLVING STORM (T.R.S.) whose diameter varies between 50 and 800 miles, 500 miles being an average. Wind speeds of over 130 knots may be experienced in the storm field which should be avoided if at all possible. After formation between 5° and 10° of latitude the storm moves westwards at 10–12 knots until reaching the tropic where it slows down before recurving eastwards and proceeding at 15–20 knots to the higher latitudes.

Tropical revolving storms do not usually cross the land, but when they do, the supply of warm moist air necessary to their existence is cut off and they tend to fill. If they return to the sea they usually deepen again.

Wind velocities vary greatly depending on the pressure gradient. Winds of 130 knots or over may be experienced. The strongest winds are usually found just abaft the trough line in the dangerous semi-circle. The velocity is greatest here because the storm tends to force itself into the sub-tropical high pressure and therefore the isobars are packed more closely together, giving a greater pressure gradient and wind than in the navigable semi-circle where the isobars are not so packed. A similar packing of the isobars takes place on the equatorial side of many temperate latitude depressions.

In the central part of the storm (the eye or vortex) the winds are quite light, the sky is usually fairly clear as well, although conditions are far from pleasant as there is usually a high confused swell.

It must be emphasised that not all storms follow such an ideal path as that outlined above and shown in the diagram on page 73. Reference to charts showing storm tracks shows what varied paths a T.R.S. can take, but if the rules for avoiding storms are followed the storm centre can almost always be avoided.

The following table shows the areas in which tropical storms are likely to be encountered, together with the most likely season and the local name. It will be noted that the T.R.S. is not found in the South Atlantic or the eastern side of the South Pacific. Reference to the charts on pages 41 and 42 will show that the ITCZ is at no time south of 5°S in these areas.

AREA	NAME	SEASON
North Atlantic Ocean Western side	Hurricane	June to November
North Pacific Ocean Eastern side	Hurricane or Cordonazo	June to November
Western side	Typhoon or Baguios	All the year but greatest frequency and intensity June to November.
South Pacific Ocean Western side	Hurricane	December to April
South Indian Ocean Eastern side	Willy-willy	December to April
Western side	Cyclone	December to April
Bay of Bengal and Arabian Sea	Cyclone	June and November but they may occur during the S.W. monsoon season.

The following terms are in common usage when reference is made to a T.R.S., some are shown in the drawings on the following page :—

PATH. The direction in which the storm is moving.
TRACK. The area which the storm centre has traversed.
STORM FIELD. The horizontal area covered by the cyclonic conditions of the storm.
SOURCE REGION. The region where the storm first forms.
VERTEX OR COD. The furthest westerly point reached by the storm centre.
EYE OF THE STORM. The storm centre.
BAR OF THE STORM. The advancing edge of the storm field.
ANGLE OF INDRAUGHT. The angle which the wind makes with the isobars.
VORTEX. The central calm of the storm.
DANGEROUS SEMI-CIRCLE. The half of the storm which lies to the right of the path in the Northern Hemisphere and to the left of the path in the Southern Hemisphere.
DANGEROUS QUADRANT. The leading portion of the dangerous semi-circle where the winds blow towards the path.

NAVIGABLE SEMI-CIRCLE. The half of the storm which lies to the left of the path in the Northern Hemisphere and to the right of the path in the Southern Hemisphere.

TROUGH LINE. A line through the centre of the storm at right angles to the path. The dividing line between falling and rising pressure.

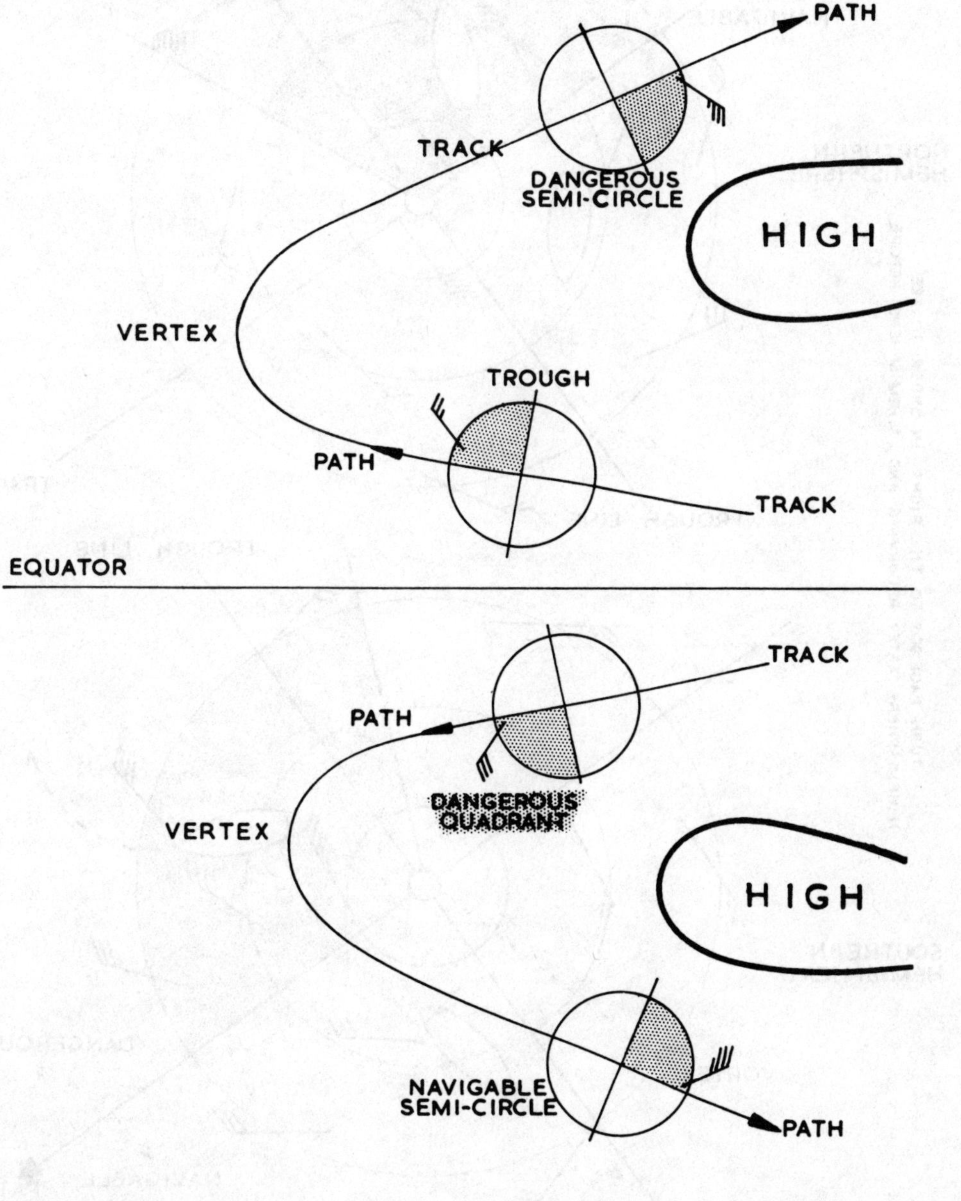

TURN PAGE 90° TO THE RIGHT IN ORDER TO SEE N. HEMISPHERE AFTER RECURVING AND S. HEMISPHERE BEFORE

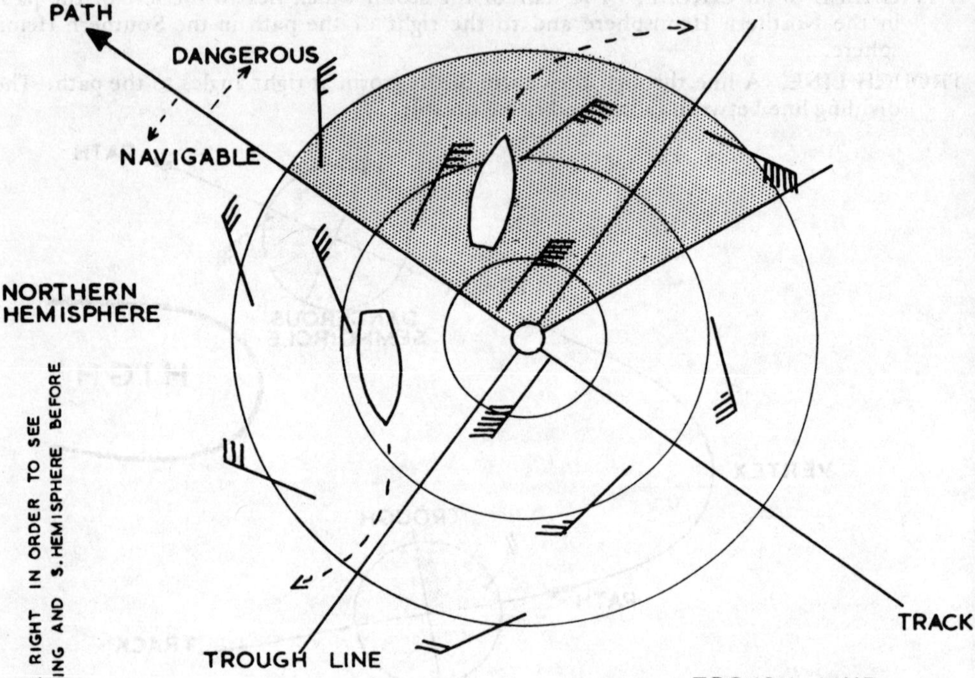

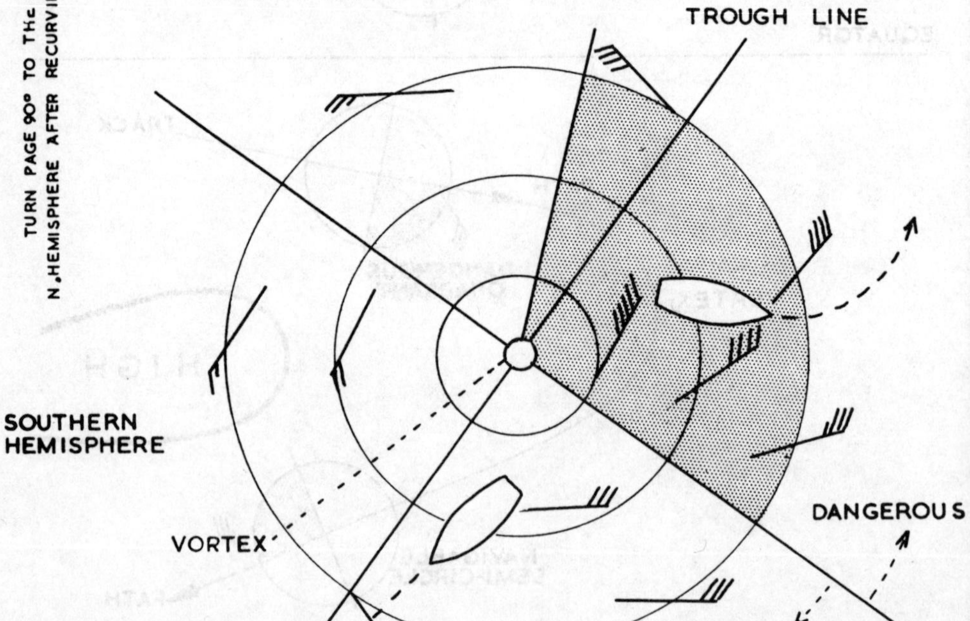

Some or all of the following signs will be evident as a TRS approaches.

1. A swell for no apparent reason. This may be felt up to 1000 miles from the storm centre.
2. Irregularity in the diurnal range of the barometer. In the tropics a barometer reading appreciably (3mb) differently from the corrected reading for that time as shown on the weather chart, should be regarded with suspicion. Frequently, a slow fall occurs between 500 and 120 miles from the centre; the diurnal range is still noticeable on the barograph trace. Between 120 and 60 miles from the centre, the diurnal range is masked and the fall is distinct. From 60 miles to the centre the fall is very rapid, and after the centre has passed the rise will be just as rapid.

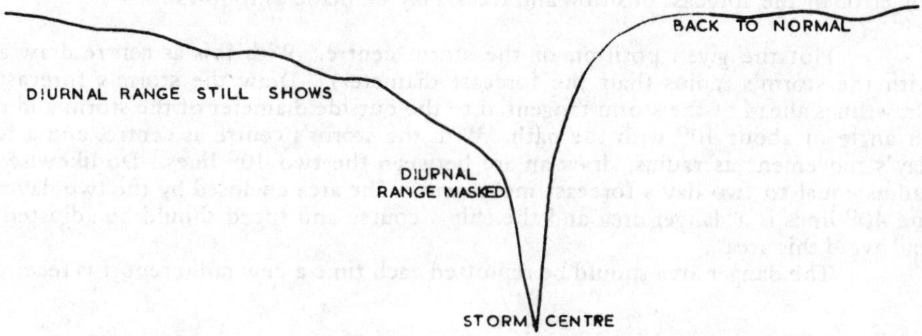

Barograph trace in vicinity of a T.R.S.

3. A change in the appearance of the sky. Cirriform cloud first appears with cirrus in bands converging towards the centre. This is followed by cirrostratus, cirrocumulus and then altocumulus with banks of black clouds on the horizon. The black cloud (nimbostratus) is called the bar of the storm. Up to this time precipitation has been sporadic but it now becomes continuous and is torrential.
4. An increase in velocity or change in direction of the trade wind, particularly with an unsteady barometer, usually indicates that a TRS is not very far away.
5. An oppressive feeling in the air.

AVOIDING THE STORM CENTRE.

If a vessel gets within the storm field she should find her position relative to the centre as soon as possible.

This can be done in the following manner; after observing the wind direction, the storm centre can be estimated as being 12 points to the right of this direction in the Northern Hemisphere when the barometer starts to fall. If the barometer has fallen 10 mb the storm centre is approximately 10 points to the right of the wind and, when the barometer has fallen a further 10 mb, the centre is about 8 points to the right of the wind. If the vessel is in the Southern Hemisphere, left should be substituted for right in the foregoing paragraph.

The storms on the western side of the South Indian Ocean have a very large angle of indraught, and in some parts the wind blows straight across the isobars towards the storm centre. In these cases the method of finding the approximate bearing of the centre fails. A freshening S.E. trade with a falling barometer indicates that the vessel is on the path of or in the dangerous semi-circle of a T.R.S.

The vessel's position in relation to the path of the storm may be found by noting the wind direction and then after an interval of time noting it once more. If the wind shifts to the right (veers) the vessel is in the right hand semi-circle, and if it shifts to the left (backs) the vessel is in the left hand semi-circle. This applies to both hemispheres. If there is no shift of wind the vessel is on the direct path if the barometer is falling, or is proceeding at the same speed and in the same direction as the storm if the barometer is unchanged. If there is doubt as to the relative movement of the ship and storm the ship should be stopped until this movement is found.

A continual check on the storm's movement must be made either by radio weather reports or, if these are not available, by estimation as outlined above.

When radio warnings of a storm's position and track are received an allowance for error in the forecast position and track may be made as follows :—

Plot the given position of the storm centre. With this as centre draw a circle with the storm's radius (half the forecast diameter). Draw the storm's forecast path. Draw lines ahead of the storm tangential to the outside diameter of the storm and making an angle of about 40° with the path. With the storm's centre as centre, and a forecast day's movement as radius, draw an arc between the two 40° lines. Do likewise with a radius equal to two day's forecast movement. The area enclosed by the two day arc and the 40° lines is a danger area and the ship's course and speed should be adjusted to try and avoid this area.

The danger area should be replotted each time a new radio report is received.

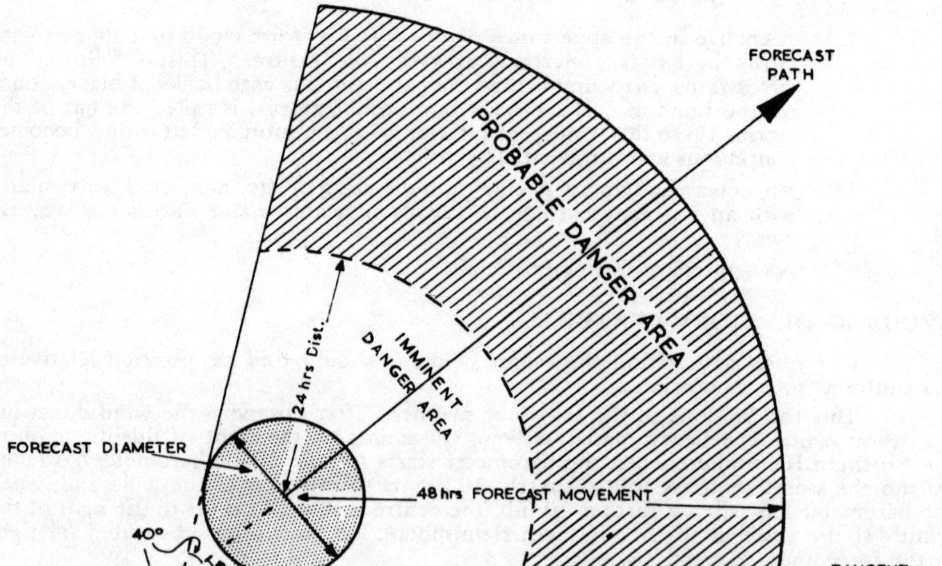

The following rules apply to vessels navigating in the vicinity of a Northern Hemisphere TRS.

If the vessel is in the dangerous semi-circle she should make as much speed as possible keeping the wind on the starboard bow. If it is impossible to make headway because of bad weather or the proximity of land, the vessel should heave to with the wind and sea in as comfortable a position as possible. To have the sea on one bow and the wind on the other bow is often a good plan, as there is quite an angle between the wind and swell.

If the vessel is in the navigable semi-circle she should run with the wind on the starboard quarter, or if there is no room to run then she should heave to as above.

If the vessel is in the direct path she should run with the wind just abaft the starboard beam into the navigable semi-circle.

If the vessel is in the Southern Hemisphere the manoeuvres outlined above should be carried out reading Port for Starboard.

If the vessel is in a hurricane or typhoon anchorage she should have both anchors down with a good scope of cable; she should also have the engines in readiness to ease the strain on the cables.

If the vessel is in an open roadstead, she will probably have a better chance of weathering the storm if out at sea, but the decision to put to sea must be taken early, otherwise the vessel may well be caught on a lee shore.

If the vessel is alongside a wharf, all hatches should be securely battened down and derricks lowered and secured. Extra moorings should be put out fore and aft. Adequate fenders should be placed between the ship and the quay. It may be advisable to lay out the anchors. The engines should be ready for instant use. In many ports it is usual for the vessel to leave the wharf and proceed to a typhoon anchorage.

In all cases a vessel should be made as stable as possible, ballast tanks filled and free surface in tanks reduced as much as possible, as during a TRS the force of wind pressing on a ship's side will cause a considerable heeling moment which could have disastrous consequences. Cargo and other moveable objects should be well secured or tommed off.

VII
Currents and Ice

Most places have a daily or twice daily rise and fall of water level known as a tide. The change in level at a port is caused by water flowing into or out of that port. This water flow is a TIDAL STREAM which is caused by the gravitational effect of the sun and moon.

OCEAN CURRENTS are not dependent upon gravitational effect but are caused by either a difference in level or a difference in density, in which case they are called STREAM CURRENTS, or by the wind blowing continuously over the surface in the same direction, in which case they are called DRIFT CURRENTS.

Stream currents may attain a rate of 3 to 4 knots. The best known examples of these are the Gulf Stream, Kuro Siwo and the surface current through the Strait of Gibraltar into the Mediterranean (whose level has been lowered by evaporation). An example of the latter type is the current which flows out of the Mediterranean below the surface (the salinity of the Mediterranean is increased by the evaporation of water).

Drift currents are usually of the order of 1 to 2 knots, the principal ones being the Southern Ocean or Great West Wind Drift and the Trade Drifts.

It will be realised that the measurement of current sets and drifts is difficult, as accurate D.R. and observed positions must be worked out by many ships on a world wide basis before being collected and finally plotted. Drift bottles or plastic envelopes are extensively used, but even when these are recovered, one cannot tell whether or not the bottle or envelope has travelled direct from the position it was thrown overboard or how long it has taken on its journey, as it may have lain on the beach for days or weeks before being found.

Since the advent of gyro compasses and wireless time signals, many of the stronger currents and certain local currents have disappeared! Many local currents still remain but these are mainly counter currents which, because of land or sea bed configurations, flow in the opposite direction to the main current.

The chart on pages 44 and 45 shows the principal currents. The warm currents are shown in red and the cold in blue. These terms are relative as the north flowing portion of the North Atlantic Drift is shown as being warm whereas the southern branch is shown as being cold.

The temperature of the sea will vary from almost 32°C (90°F) in the equatorial regions to -2°C (28½°F) at the edge of the ice in the polar regions.

If the temperature of the water is -2°C for a considerable depth ice will start to form at the surface. Navigation will not be impaired by thin ice, but unless specially strengthened for navigation in ice, no attempt should be made to force a ship through the thicker ice. However, in many areas where this occurs, icebreaker assistance is available. Due attention should be paid to broadcast warnings of ice conditions in the area the vessel is to transit. Broadcast messages frequently refer to the following ice terms, which are given in the order in which they generally form.

SLUDGE OR SLUSH. The initial stages in the freezing of sea water, when its consistency is gluey or soapy.

BRASH. Small fragments and rounded nodules; the wreckage of other forms of ice.

PANCAKE ICE. Small pieces of new ice, approximately circular and with raised rims.

YOUNG ICE. Newly formed ice.

BAY ICE. The young ice which first forms on the sea in autumn, and is of sufficient thickness to impede or prevent navigation.

PACK ICE. Term used in a wide sense to include any area of sea ice, other than fast ice, no matter what form it takes or how disposed.

FLOE. An area of ice, other than fast ice, whose limits are within sight.

FIELD ICE. Area of pack ice of such extent that its limits cannot be seen from the masthead.

LEVEL ICE. All unhummocked ice, no matter of what age or thickness.

HUMMOCK. A ridge or elevation on a floe due to pressure.

PRESSURE RIDGE. Hummocked ice where floes have been pressed together and broken against each other.

BERGY-BITS. Medium sized pieces of glacier ice, or of hummocky pack, washed clear of snow. Typical bergy-bits have been described as about the size of a cottage.

GROWLERS. Smaller pieces of ice than bergy-bits, appearing greenish in colour, because barely showing above water.

ROTTEN ICE. Floes which have become much honeycombed in the process of melting.

FAST ICE. Sea ice which remains fast in the position of growth throughout the winter, and sometimes even during the ensuing summer.

LAND ICE. Ice attached to the shore, within which there is no channel.

Large masses of land ice which become detached are known as ICEBERGS. In the Northern Hemisphere these bergs are "calved" from the glaciers on the East and West Coasts of Greenland. Many of these bergs get stuck fast in the pack ice on the East Greenland Coasts, but those that do not are taken south by the East Greenland current to Cape Farewell whence they are carried into the Davis Strait where they melt. Other large bergs calved on the West Greenland coast are carried northward to Baffin Bay before returning south in the Labrador current to reach the shipping lanes off the Banks of Newfoundland. Due to the short summer the bergs usually reach the shipping lanes the year after they have calved, having been frozen in the pack ice for their first winter. The maximum size of a northern berg is unlikely to exceed ¼ mile in length. Its height may be up to 90mm (270 feet). It should be noted that about $^8/_9$ths of the iceberg is submerged and the underwater portion may extend laterally quite a way from the portion above water. The eastern limit of the area in which icebergs may be found is about 40°W and the southern limit about 40°N.

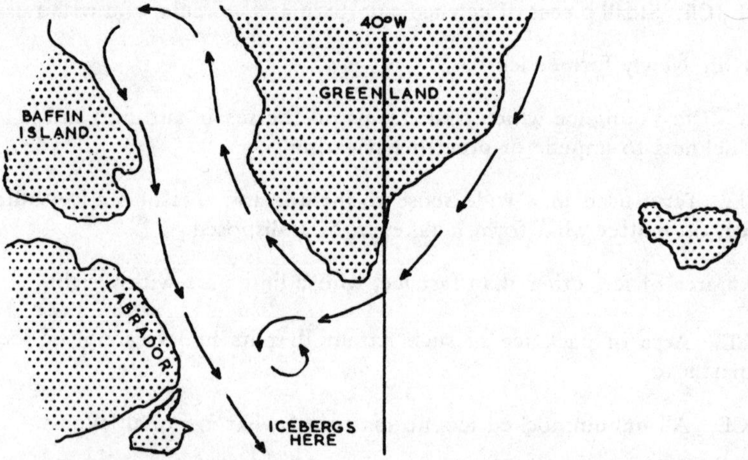

In the Southern Hemisphere the bergs are calved from the ice foot and tend to be flat topped or tabular up to 60m (180 feet) high. Their length is considerably greater than the Northern Hemisphere bergs. Bergs up to 70 miles long and 10 miles wide have been reported. The bergs are rarely encountered to the northward of 40°S and, as there is not much traffic in the higher Southern latitudes, less importance is attached to Southern bergs than to their Northern counterparts.

INDICATIONS OF ICE.

1. Ice blink by day and night. A yellowish white light reflected from the ice onto the sky near the horizon.
2. There is often a wall of thick fog at the ice edge.
3. A rapid fall in the sea temperature to below freezing.
4. Herds of seals or flocks of birds far from land.
5. Absence of sea or swell in a fresh breeze is an indication of ice or land to windward.
6. Noise of the ice cracking or falling into the sea may be heard.
7. Radar. The echoes will depend on the configuration of the ice. Also in the case of a berg, on the amount of Moraine (glacial material) in it. It should be noted that radar is not infallible in picking up ice.
8. On calm days echoes may be heard from higher ice.

NAVIGATING IN ICE.

Vessels navigating in areas where icebergs may be encountered should go at a moderate speed keeping a good lookout. If in fog, as frequently occurs on the Banks of Newfoundland, extra special care must be taken.

If navigating through pack ice the lanes and leads should be followed. The leads may be seen showing up as dark lines or patches on the ice blink. The greatest care must be exercised to prevent the vessel becoming "nipped" in the ice. This may occur if the ice is hummocking, and no attempt should be made to enter or cross ice which is forming these pressure ridges.

If collision with a berg cannot be avoided then every endeavour should be made to collide with it head on.

INTERNATIONAL ICE PATROL SERVICE.

This is financed by the countries whose shipping uses the Northern Atlantic, but is maintained by United States Coast Guard vessels whose call sign whilst on ice patrol is NIDK. The object of this patrol is to locate by air and surface scouting, and radio information from all sources, the icebergs and field ice nearest to and menacing the North Atlantic Lane Routes. The Southerly limits of the ice will be determined and contact maintained with the ice as it moves south by a continuous surface patrol.

The patrol continues throughout the ice season from approximately 1st March to 1st July or later or earlier if conditions warrant it.

Ships sighting ice should report its position to the ice patrol vessel, or to the headquarters of the service at Argentia Newfoundland, call sign NIK.

NORTH ATLANTIC LANE ROUTES.

Vessels trading accross the North Atlantic are expected to keep to the track specified for the time of year. There are seven of these tracks, each track having Westbound and Eastbound routes.

The Transatlantic Track Association comprises the principal North Atlantic Shipowners and decisions made by them are communicated by means of Notices to Mariners and the shipping press.

VIII
Weather Charts and Routeing

SYNOPTIC CHARTS.

In addition to the Ocean Weather Ships, which are stationed permanently at selected positions in the North Atlantic, there are Merchant Ships observing and radioing weather information every six hours. From this information and from information supplied by land stations all over the Northern Hemisphere, the Meteorological Office prepares weather charts.

Selected and Supplementary Ships radio their position, the wind direction and speed, the pressure and temperature, the amount of cloud and cloud types, the visibility, and the past and present weather. Selected ships also radio information about the sea temperature, the dewpoint temperature, waves and the ship's course and speed. The information to be radioed is coded in the form given in the "Decode for Use of Shipping" (Met.O.509) to which reference should be made for full details of weather messages.

Weather forecasts for shipping are broadcast by the BBC on the 1500 metre wavelength programme. The forecast areas lie to the east of 40°W and between 35°N and 60°N, the areas around the coasts of the British Isles being smaller than those in the Atlantic.

An Atlantic Weather Bulletin for Shipping from "Bracknell Weather" to "All Ships" is broadcast from Portishead radio. This broadcast is in six parts, parts I, II, and III consisting of storm warnings, a synopsis of weather conditions and weather forecasts for areas from 35°N to 65°N between 15°W and 40°W. These are all given in plain language and are broadcast at 0930 and 2130 G.M.T.

Parts V and VI are also broadcast at 0930 and 2130 G.M.T., and consist of a selection of ship and shore reports. These are broadcast in code and besides the ships' positions and station identification numbers the following information is included:— the wind direction and speed in knots, the barometric pressure, the air temperature, cloud amount, visibility, and past and present weather. This information can be plotted in the form of a station model, as shown below, at the approximate position on a chart. The ideal chart for plotting weather observations is Metform 1258.

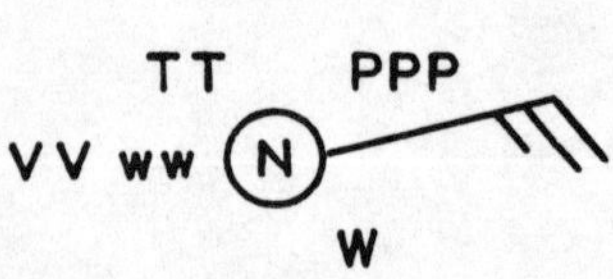

PPP is the pressure; the initial 9 or 10 is usually omitted

TT is the air temperature

VV is the code figure for the visibility

N is the Beaufort symbol for the cloud amount

ww is the Beaufort symbol for the present weather

W is the Beaufort symbol for the past weather

The line with the feathers represents the wind direction and speed. The arrow flies with the wind and a speed of 5 knots is represented by a short feather, 10 knots by a long feather, 15 knots by a short and a long feather, and so forth. The feathers always point towards the low, so that in the northern hemisphere the feathers are always on the observer's right hand when he faces the wind. In the example a wind of 075°T 25 knots has been plotted.

Part IV of the bulletin is broadcast at 1130 GMT and consists of a weather analysis. Contained herein is sufficient information to construct a weather chart. Metform 1258 should be used for the plotting of the following given data:—

The positions of high, low or other pressure systems.

The type of front and its positions

The positions of selected isobars.

When the information given in parts IV, V and VI of the Atlantic Weather Bulletin has been plotted on Metform 1258, the chart should be completed by drawing in isobars at 4 millibar intervals between those given in Part IV.

Bearing in mind the rules for the movement of depressions given on page 62 and the pressure characteristic (broadcast in Part IV) a weather forecast may be made for a selected position for the next few hours.

If examination candidates have to construct a weather chart from coded information similar to that broadcast in the Atlantic Weather Bulletin, the Decode Book Met. O.509 will be supplied but, nevertheless, practice is required in plotting the data and drawing in the isobars if the chart is to be completed within the time available. It is suggested that this practice could be done most profitably at sea, using actual broadcast information.

It is also suggested that candidates who are pressed for time should only put in the wind, temperature and pressure on the station model.

It may be found occasionally that some information in Part IV, V or VI does not correspond with the general synoptic situation. This should be disregarded, as there may have been an error in receiving or decoding the message.

FACSIMILE WEATHER CHARTS.

Vessels fitted with facsimile plotters can receive signals by radio which will produce, on a moving roll of paper, a weather chart in the same form as those drawn by the Meteorological Office. This removes any possibility of mistakes in reading, decoding or plotting weather information sent out by code as in the Atlantic Bulletin.

Impressions of black or white signals are made on a roll of paper as it passes an electronic marker pen. Different types of transmissions require different drum speeds and in order that the chart reproduction is perfect the receiver must be kept correctly tuned and the drum speed also correct throughout the transmission. Any spots which are too dark or insufficiently dark cannot be overprinted by the machine as the transmission takes several minutes and the chart cannot be fed back into the machine. If a bad chart is produced a further chart will have to be produced at the next transmission.

Full details of the scale and type of the map projection, the drum speed, the type of chart (surface analysis, surface prognosis i.e. forecast, wave analysis, wave prognosis, sea ice observations etc.), the transmitting stations with wavelengths and times are given in Volume III of the List of Radio Signals.

WEATHER ROUTEING.

This is carried out by private companies in the U.S.A. and by the British Meteorological Office in the U.K. A small charge is made for the service.

During the comparatively short period that it has been in operation many hundreds of crossings of the North Atlantic have been made by vessels using this service. Whilst it is impossible to give the time saved or reduction in heavy weather damage for individual passages, the average time taken by weather-routed vessels is less than that of similar vessels not so routed and the damage caused by heavy weather during a routed period is significantly less than that during a similar period when not routed. The following description of weather routeing is taken from the Meteorological Office leaflet on weather routeing with the permission of the Controller of H.M. Stationery Office and the Director of the Meteorological Office.

To enable ships to make the best of the weather the British Meteorological Office has recently introduced a weather routeing service for vessels trading on the North Atlantic. Voyages in both east-west and west-east directions are covered and ships of any nation may participate.

Each ship is given individual advice on the best course to steer in order to avoid heavy weather as far as possible. The Master still remains responsible for deciding his course, but experience has shown that the use of weather-routeing advice can offer the following advantages.

a. Higher speeds

b. Quicker crossings

c. Better punctuality

d. Greater comfort for passengers and crew

e. Less damage to ships

f. Less damage to cargoes

g. More opportunities for maintenance at sea.

Just before sailing the Master contacts the Meteorological Office at Bracknell where the Central Forecast Office gives advice covering at least the next 48 hours. This can be very important. For example, should a ship leaving Liverpool go north around Ireland, or will it pay to go the longer way via Fastnet? The shortest route is not always the quickest.

This initial advice is followed by further advisory messages transmitted to the ship by Portishead Radio. These are normally sent every 48 hours, but messages can be sent more frequently – every 12 hours if necessary. These messages advise the course to follow and, if required, a special weather forecast for 48 hours and forecasts of wind and sea can also be provided.

The Central Forecast Office at Bracknell receives a constant flow of meteorological data from the North Atlantic and adjacent areas. These data are fed into a high speed electronic computer which produces regular forecasts for the North Atlantic for 2–3 days ahead. The computer converts the predicted wind field into wave heights and makes allowance for swell. Comprehensive information about currents and ice movements is also available. In order to know as much as possible about the performance of an individual ship in varying wave and weather conditions, the Office asks in advance for all the necessary details about each ship and each voyage, including, for example, the size, draught and loading of the vessel and any special characteristics which might affect her performance on a particular crossing.

Armed with all this information, expert weather forecasters, aided by nautical officers with long ocean-going experience, calculate the best course for the ship to follow during the next 48 hours to enable it to reach its destination in the shortest time and with the minimum of bad weather. Sometimes, of course, bad weather is too widespread to avoid completely, but practical experience has proved that over a few voyages weather routeing soon pays for itself in terms of time saved and damage avoided.

Page 86

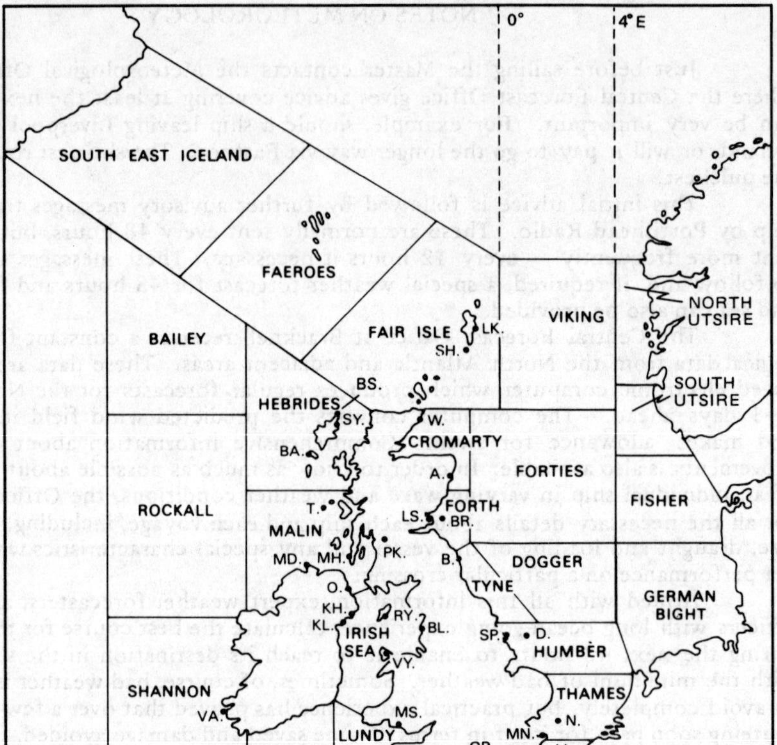

Sea areas and coastal stations used in marine forecasts prepared by the Meteorological Office at Bracknell and broadcast by the BBC and Coast Radio Stations, incorporating changes to areas made in 1984. Weather within an estuary, and particularly visibility, may often be expected to differ from that forecast for the coastal sea areas within which the estuary lies, though this will not necessarily be mentioned in the forecast.

Index

ABSOLUTE HUMIDITY 10
Absolute temperature 9
Adiabatic temperature changes 18
Advection fog 30
Air Masses 54,55
Anabatic wind 51
Anemometers 12
Aneroid Barometer 6,7
Angle of Indraught 37
Anticyclone 53
Anti-trades 43
Arctic smoke 30
Atmosphere 15
Aureole 31

BAILLIE WIND ROSE 40
Barograph 8
Barometer 4
Barometer errors 5
Beaufort wind scale 38
Bergeron process 25
Bishop's Ring 31
Black frost 31
Buys Ballot's Law 34

CAPACITY 5
Capillarity 5
Celsius 9
Centigrade 9
Cloud 23,24
Col 65
Conditional instability 20
Conduction 17
Convection 17
Coriolis 33
Corona 31
Cumulonimbus cloud 26
Current chart 44,45
Currents 78
Cyclostrophic force 36

DEPRESSIONS 56,63
Dew 29
Dew point 10
Diurnal range 3
Dry Adiabatic Lapse Rate 18

EASTERLY TYPE 68

FACSIMILE CHARTS 84
Fahrenheit 9
Fiducial temperature 5
Fog 29
Föhn wind 52
Forecast Areas 86
Friction 37
Frontogenesis 61
Frontolysis 61
Fronts 55,56
Frost 29

GEOSTROPHIC WIND 34
Geostrophic Wind Scale 35
Glazed frost 31
Gold slide 6
Gradient wind 36

HAIL 26
Halo 31
Haze 29
Hoar frost 29
Humidity 10
Hydrometer 13
Hygrometer 11

ICE 79,80
Ice Patrol Service 81
Insolation 16
Inversion 21
Isallobar 3
Isobar 3
Isobaric forms 66
Isohaline 14
Isohel 13
Isohyet 13
Isoneph 14
Isotherm 9
KATABATIC WIND 51

Kew barometer 4

LAND AND SEA BREEZES 43
Lapse rates 18
Latent heat 15
Line squall 64
Local winds 49,50

MAXIMUM THERMOMETER 10
Mediterranean winds 49
Mercurial barometer 4
Millibar 3
Minimum thermometer 10
Mirage 32
Mist 29
Mock sun 31
Monsoons 46,47
Movement of depressions 62

NEPHOSCOPE 14
N.E. Monsoon 46
N.E. winds of Brazil 48
N.W. Monsoon 46
Neutral air 20
Northerly Type 67

OCCLUSION 56
Optical phenomena 31
Orographic fog 30

PLANETARY CIRCULATION 40
Precipitation 25
Precision Aneroid Barometer 7
Pressure 3
Pressure Gradient 3
Prevailing winds, N. Summer 42
Prevailing winds, N. winter 41
Psychrometer 11
Pumping 5

QUASI-STATIONARY FRONT 55

RADIATION 17,29
Radio-sonde 14
Rain 25
Rainbow 32
Raingauge 13
Réamur 9
Relative Humidity 10

Rime 31
Ridge 54

ST. ELMO'S FIRE 28
Salinometer 14
Saturated Adiabatic Lapse Rate 18
Sea breeze 43
Sea fog 30
Sea smoke 30
Secondary Depression 61
Sixes Thermometer 11
Sleet 26
Snow 26
Solar Constant 16
Source region 54
Southerly type 69
S.W. Monsoon 47
S.W. winds of Africa 48
Specific heat 9
Stable air 19
Standard Temperature 5
Storm warnings 39
Straight isobars 65
Sunshine recorder 13
Synoptic charts 82

TEMPERATURE 9,10
Tephigram 22
Thermograph 11
Thermometers 10
Thunderstorm 27
Tornado 64
Trade wind limits 43
Tropical Revolving Storms 71-78
Troposphere 15
Trough 63
Turbulence 17

UNSTABLE AIR 19

WATERSPOUT 64
Wave depression 61
Weather charts 82
Weather Routeing 84
Westerly type 70
Wind 33
Wind and Pressure – summer 42
Wind and Pressure – winter 41
Wind rose 39,40

FURTHER READING

METEOROLOGY AT SEA

Ray Sanderson

Practical forecasting and the weather and sea conditions that concern mariners, explained by a professional marine and aviation forecaster who has worked in Europe and the British Isles. The author is responsible for the preparation of marine forecasts at the Bracknell Meteorological Office and is also an experienced seaman.

Contents include:

—Factors producing winds, precipitation, altered visibility and ice.

—Pressure features and the movement of weather systems in both Northern and Southern Hemispheres.

—Signs of approaching bad weather.

—Forecasting, and the use of forecasts and weather maps.

—Tropical revolving storms—hurricanes, typhoons and cyclones: their development and movement, signs of approach, and avoidance tactics for ships.

—Waves, wave effects, ocean currents, sea ice and iceing.

—Weather in the main Atlantic sea areas, the Baltic and Mediterranean, and around the British Isles.

227 pages 170 diagrams and tables
19 photos ISBN 0 540 07405 5